William Smith Clark

Observations on the Phenomena of Plant Life

a paper presented to the Massachusetts Board of Agriculture

William Smith Clark

Observations on the Phenomena of Plant Life
a paper presented to the Massachusetts Board of Agriculture

ISBN/EAN: 9783337095291

Printed in Europe, USA, Canada, Australia, Japan

Cover: Foto ©berggeist007 / pixelio.de

More available books at **www.hansebooks.com**

[From the 22d Annual Report of the Secretary of the Mass. State Board of Agriculture.]

OBSERVATIONS

ON

THE PHENOMENA OF PLANT LIFE.

A PAPER

PRESENTED TO THE

MASSACHUSETTS BOARD OF AGRICULTURE,

By W. S. CLARK,

PRESIDENT OF THE MASSACHUSETTS AGRICULTURAL COLLEGE.

BOSTON:
WRIGHT & POTTER, STATE PRINTERS,
79 MILK STREET (CORNER OF FEDERAL).
1875.

PHENOMENA OF PLANT-LIFE.

The observations concerning "The Circulation of Sap in Plants," which I had the honor of presenting before the Board of Agriculture at their last country meeting, were so kindly received at the time, and awakened so much interest after their publication, that I have found it impossible to refrain from further investigations upon the phenomena of plant-life. Among the subjects to which special attention has been directed during the year, the following may be enumerated, viz.:

First. The structure, composition and arrangement of the winter-buds of hardy trees and shrubs. Specimens for study were collected, in January and February last, from one hundred and forty species, and some facts of interest recorded.

Second. The percentage of water to be found in the branches and roots of trees during their annual period of repose, as well as when in active growth.

Third. The phenomena and causes of the flow of sap from wounds in trees when denuded of their foliage, as well as the flow from the stumps of woody and herbaceous plants when cut near the ground in summer. In connection with this subject, an attempt has been made to determine what species flow, how rapidly and copiously, and under what circumstances.

The pressure exerted by the sap exuded from detached roots of trees under ground, as well as that exhibited upon gauges placed at different elevations from the earth, has also been very carefully observed upon a number of species.

The facts determined are even more remarkable than were noticed last year, and are particularly important in the case of the sugar-maple.

Fourth. The structure and functions of the bark of exogenous trees, with special reference to the circulation of sap, the formation of wood and the effects of girdling,—concerning all which points many experiments have been undertaken with satisfactory results.

Fifth. An attempt has been made to measure the expansive force of growing vegetable tissue, and in connection with this experiment numerous other interesting observations have been reached.

These investigations have been instituted by myself; but in carrying them out, I have enjoyed the valuable, and, in many cases, indispensable, assistance of gentlemen connected with the Agricultural College, either as officers or students. Due credit will be given to each in stating the results of his work.

To succeed as an original investigator in science, one must possess some of the noblest qualities of mind and heart. He must be absolutely and accurately honest, and in his methods of demonstration there must be no guess-work. He has need of a patience which is inexhaustible, a zeal and energy which never flag, and a spirit of devotion to his work which utterly ignores self as separated from the object to be accomplished. He must also have a well-disciplined mind, and skill in the use of books and apparatus. To produce such men, who shall, at the same time, be familiar with all the great principles and problems of agriculture, is the highest possible achievement of our College. One such graduate will do more for the advancement of the art, and the honor of the profession and the benefit of mankind, than would a host of mere farm-apprentices possessed only of manual skill and a knowledge of simple, routine practice, however well adapted to any particular locality or style of farming.

I am well aware that there are persons who hold a respectable position in society, and yet are so ignorant as to regard with contempt all efforts at scientific research. They ridicule the attachment of gauges to trees, and the harnessing of squashes, and the microscopic and chemical analysis of plants, as of no earthly use, except, perhaps, to gratify an idle curiosity. But how shall agriculture be improved without the application to it of the principles of science ; and how shall these be applied unless they are discovered; and how shall they be known, if they are not sought? In no way can the wealth of the world be increased so surely as by the liberal endowment of institutions for the special purpose of securing experiments in all departments of science which have a direct connection with agriculture, especially in chemistry and in animal and vegetable physiology. When we consider that, to observe the transit of Venus during the present month, expeditions have been sent to different parts of the earth, at a cost of more than a million of dollars, we may, at least, hope that scientific observations upon things nearer home, and having more to do with every-day life, will soon be appreciated and supported.

We are told that when the illustrious scientist, Faraday, who devoted his life to original research, was asked by some practical individual what was the use of one of his famous discoveries, he answered him by propounding another equally pertinent question, namely, "What is the use of a baby?" The possible results are in both cases of transcendent moment, but in neither can they be foretold. It is enough to know that every new truth is an open door to some further discovery and to some useful invention.

It has been well said that it is comparatively easy to know something about everything, but very difficult to learn everything about anything. Remembering that we are enveloped by inexplicable mysteries, and that abundant material for investigation lies everywhere about us, we have attempted to study that most familiar plant, *the squash*,—and the results have far surpassed our most sanguine expectations.

The particular species selected for observation is named *Cucurbita maxima*, and the variety is called, by Gregory, the mammoth yellow Chili. It is said to be a native of the Levant, and to have been introduced into England in 1547. It is sometimes called the French pumpkin, and its fruit readily attains a weight of one hundred and fifty pounds. One has been grown in England which weighed two hundred and forty-six pounds.

Squashes indigenous to tropical America were cultivated by the Indians long before the occupation of this continent by the whites.

The *Cucurbitaceæ* are a small, but very useful order of the vegetable kingdom, numbering about three hundred and fifty species, which are chiefly natives of warm regions. The most valuable species are the squash, the pumpkin, the cucumber, the water-melon, the musk-melon and the gourd, of all which there are numerous varieties.

These plants are generally herbaceous, and trailing or climbing by means of tendrils. Their stems, leaf-stalks, tendrils and fruits are often hollow, and their tissues very soft and succulent.

The flowers are usually large, and either yellow or white, and of two or three sorts on the same plant. The fruit is commonly a pepo, the structure of which is familiar to all.

The following considerations suggested the idea of experimenting with the mammoth squash:

First. It is a well-known fact that beans, acorns and other seeds often lift comparatively heavy masses of earth in forcing their way up to the light in the process of germination.

Second. We have all heard how common mushrooms have displaced flagging-stones, many years since, in Basingstoke, and, more recently, in Worcester, England. In the latter case, only a few weeks ago, a gentleman noticing that a stone in the walk near his residence had been disturbed, went for the police, under the impression that burglars were preparing some plot against him. Upon turning up the stone, which weighed eighty pounds, the rogues were discovered in the shape of three giant mushrooms.

Third. Bricks and stones are often displaced by the growth of the roots of shade-trees in streets. Cellar and other walls are frequently injured in a similar way.

Fourth. There is a common belief that the growing roots of trees frequently rend asunder rocks on which they stand, by penetrating and expanding within their crevices.

Having never heard of any attempt to measure the expansive force of a growing plant, we determined to experiment in this direction.

We were surprised, last year, in testing the pressure exerted by the sap of various trees, to find that a black birch-root detached from the tree, was able to force water to the height of eighty-six feet. We were therefore somewhat prepared for an exhibition of considerable power, but the results of our trials have, nevertheless, been most astonishing.

At first, we thought of trying the expansive force of some small, hard, green fruit, such as a hickory-nut or a pear, but the expansion was so slow, and the attachment of the fruit to the tree so fragile, that this idea was abandoned. The *squash*, growing on the ground with great rapidity and to an enormous size, seemed, on the whole, the best fruit for the experiment.

Accordingly, seeds having been obtained from Mr. J. J. H. Gregory, of Marblehead, they were planted on the first of July in one of the propagating pits of the Durfee Plant-House, where the temperature and moisture could be easily controlled. A rich bed of compost from a spent hot-bed was prepared, which was four feet wide, fifty feet long, and about six inches in depth. Here, under the fostering care of Prof. Maynard, the seeds germinated, the vine grew vigorously, and the squash lifted in a most satisfactory manner.

Never before has the development of a squash been observed more critically, or by a greater number of people. Many thousands of men, women and children, from all classes of society, and of various nationalities, and from all quarters of the earth, visited it. Mr. D. P. Penhallow watched with it several days and nights, making hourly observations. Prof. H. W. Parker was moved to write a poem about it, and Prof. J. H. Seelye declared that he positively stood in awe of it.

Vegetable growth consists in the development of the several parts of a plant, according to a definite, predetermined plan as regards the form, size and other characteristics of each species. It results from the activity of a certain peculiar inherent force, called life. Under the influence of this force, stimulated to action by heat and light, plants absorb, digest and assimilate mineral matter, converting it into the various organic substances which enter into their composition. Examined under the microscope, all parts of plants are found to consist primarily of closed cells, cohering into masses of various forms and containing protoplasm.

Growth is caused by the increase of cells in number and in size. In a growing portion of a plant, as at the tip of the stem, the first-formed cells are subdivided, and then the subdivisions enlarge to the normal size, and this process goes on while growth continues. All vegetable material is primarily formed in the leaves or green parts of ordinary plants, and, by a vital process of circulation, is transferred in a liquid form to its proper destination.

The seed is a minute plant, consisting of a radical or little root, a terminal bud called the plumule, and one or more seed-leaves, all snugly packed away in a shell for safe keeping during transportation. In order that the sprouting plantlet may be able to get hold of the earth for its water and mineral supplies, and have substance enough to reach up into the light and air where it is to find its future carbon, the seed-leaves, or cotyledons, are formed of very condensed and complex materials,—such as oil, sugar, starch and albuminoids. The requisite conditions of germination for a sound, living seed are air, water, and a moderate degree of heat. The time intervening between the planting of a seed and the appearance of the root varies from a few hours to many months. It may be hastened in some cases by scalding the seed for a few minutes in hot water, or by the judicious use of a solution of camphor, sal-ammoniac, or oxalic acid. The cotyledons of the squash-seed are pushed up into the air, where they expand and thicken, assume a green color, and for a time perform the functions of true leaves.

The root is the first part of a plant to grow, and develops downward, as if affected by the force of gravity. Light neither hurts nor helps the root, but water is essential to its life, and for this it penetrates the soil in every direction. It is the special function of the root to absorb and furnish to the rest of the plant, water, nitrogenous matter, and such soluble minerals as each species requires for its use. For this purpose it is admirably adapted by its peculiar structure, substance and mode of increase. The older portion of roots serves to sustain the stem and hold it in place, and also acts as a reservoir of supplies to the plant. The younger roots usually branch off in an irregular manner, and elongate by the multiplication of cells near their extremities. The tips of roots are usually very minute fibres of exceedingly delicate tissue, which insinuate themselves into the pores of the soil, and then, by the expansive power of growth, enlarge these capillary channels to any required size.

Roots of ordinary plants grow most freely in a loose, well-drained soil, containing the essential elements of plant-food in a soluble form. They absorb their water from the surface of the molecules of the soil, to which they attach themselves by very minute, cellular papillæ, called root-hairs. These hairs are much more numerous in a soil moderately dry than in one which is wet and heavy. The most vigorous plants have the largest number and greatest extent of roots. Hence the importance of deep and thorough tillage in preparing the ground for crops. The growth of a plant depends chiefly upon the amount of water which is exhaled by its leaves, and this necessarily depends upon the supply furnished by the the roots. The folly of ploughing between rows of corn, or other plants, after their roots have spread widely through the soil, is self-evident. Prof. L. B. Arnold says he has known the maturing of a corn-crop postponed ten days by ploughing it at the last hoeing.

The penetrating power and tendency of roots is well illustrated in the case of an apple-tree on the College farm, which forced its roots down through a mass of coarse gravel eight feet, to obtain a supply of water. The stones were about the size of hens' eggs, and so closely packed by the waters of the drift period which deposited them, that the cylindrical form of the roots was entirely destroyed. The growing tissues pressed themselves into every crevice so as actually to surround and enclose the adjoining pebbles. (Fig. 17.) A similar root of an elm was recently dug up in Westfield, Mass., and presented to the College museum by Mr. B. H. Averell. Prof. Stockbridge, last fall, washed out a root of common clover, one year old, growing in the alluvial soil near the Connecticut River, and found that it descended perpendicularly to the depth of eight feet. Mr. Mechi, of Tiptree Hall, England, tells us that the reason clover is usually so short-lived, is the fact that the lower roots are either unable to penetrate the subsoil or to find in it the requisite supplies of food. He also states that his neighbor, Mr. Dixon, of Riven Hall, dug a parsnip which measured thirteen feet six inches in length, but, unfortunately, was broken at that depth.

The roots of lucerne often penetrate to the depth of more than twenty feet, while the tap-roots of trees, continuing to grow for a long period, descend still further. A British officer in India reports that the root of a leguminous tree—the *Prosopis spicigera*—is often dug for economical purposes, and that he has seen an excavation sixty-nine feet deep made for such a root without reaching its lower extremity. The roots of trees are well known to extend in a horizontal direction to surprising distances, and to exert a very deleterious influence on crops in their vicinity. The living roots of an elm, in Amherst, were found in abundance at a distance of seventy-five feet from the trunk, which was just the height of the tree. It has recently been stated in "The Field," an English paper, that the roots of an elm were found to obstruct a tile-drain which was four hundred and fifty feet from the tree.

But our squash-vine affords the most astonishing demonstration of all that has been said about root-development. Growing under the most favorable circumstances, the roots attained a number and an aggregate length almost incredible. The primary root from the seed, after penetrating the earth about four inches, terminated abruptly and threw out adventitious branches in all directions. In order to obtain an accurate knowledge of their development, the entire bed occupied by them was saturated with water, and, after fifteen hours, numerous holes were bored through the plank-bottom, and the earth thus washed away. After many hours of most patient labor, the entire system of roots was cleaned and spread out upon the floor of a large room, where they were carefully measured. The main branches extended from twelve to fifteen feet, and their total length, including branches, was more than two thousand feet. At every node, or joint, of the vine, was also produced a root. One of these nodal roots was washed out and found to be four feet long, and to have four hundred and eighty branches, averaging, with their branchlets, a length of thirty inches, making a total of more than twelve hundred feet. As there were seventy nodal roots, there must have been more than fifteen miles in length on the entire vine. There were certainly more than eighty thousand feet; and of these, fifty thousand feet must have

PHENOMENA OF PLANT-LIFE. 11

been produced at the rate of one thousand feet or more per day.

Now, it has been said, that corn may be heard to grow in a still, warm night, and it has been proved that a root of corn will elongate one inch in fifteen minutes. But here are twelve thousand inches of increase in twenty-four hours. What lively times in the soil, where such vital force is at work! The wonder is, we do not hear the building of these roots as it goes on.

But in addition to the movements caused by the increase of the roots among the particles of the soil, we should remember that solution, chemical affinity, diffusion and capillarity, as well as the absorption of the feeding rootlets, are incessantly at work beneath the surface of the silent earth. With what amazement should we behold the development of a crop upon a fertile field, if we could but see with our eyes the things which are known to transpire!

Let us next consider some peculiarities of plant-growth which were exhibited in the development of the squash-vine, with its appendicular organs—the leaves and the tendrils, and its reproductive organs—the flowers and the fruit.

The peculiar feature of the *vegetable stem* is the *bud*, by which it is always terminated, even in the seed. A bud is an aggregation of delicate cells, filled with protoplasm, and endowed with special vitality. Sometimes it is very minute and simple in structure, and sometimes large and complicated. As the stem elongates, it usually produces, at regular intervals, leaves, in the axils of which are formed buds, which, in growing, become the terminal buds of branches. The places where leaves are borne are called *nodes*, and the spaces on the stem between these are styled *internodes*. Every species of plant has a definite law for the arrangement of its leaves. Our squash produced one leaf at each node, and all the leaves were arranged in two rows on opposite sides of the stem. The vital force in the tip of the vine was very active and vigorous, and displayed its power in the constant organization of new nodes. Thus, when we examined the terminal inch of the vine, we found no less than twenty-five young leaves, and in the axils of these twenty-five flowers, including five young squashes, twenty-five branching tendrils, and twenty-five buds for lateral branches. These were continually reproduced, so that when the vine was growing nine inches a day, as well as after it had developed one hundred nodes, the number was always about the same. All parts of the vine and its appendages increased with marked uniformity. Back of the first inch, which may be regarded as the terminal bud, about six nodes were developing at the same time. The growth was most rapid in the terminal portion of each node, and the leaves were not modified particularly in form during the period of development. The lengthening of the vine proceeded somewhat irregularly, varying from nothing to nine-sixteenths of an inch per hour. It was usually less between midnight and sunrise than at other hours.

The longest growth of the main vine in twenty-four hours was observed August 15th and 16th, from 7 A. M. to 7 A. M., and amounted to nine inches. The laterals were removed when two or three feet in length. The total extent of the main vine was fifty-two feet, and the number of nodes was one hundred. At each node of the fully-developed vine were found a large leaf; a long, branching tendril, resembling the veins of a leaf, without the intervening cellular tissue; a staminate flower on a long stalk, or a pistillate flower on a short stalk; a lateral branch, and, on the under side of the vine, a long, branching root. The function of this root was evidently to supply water to the leaf above it, and its development, of course, depended chiefly upon the nutrient material elaborated by this leaf. These nodal roots not only furnished a much larger feeding-ground for the plant, but saved an immense amount of mechanical work in reducing the distances through which the crude and elaborated saps must be carried.

The largest leaves of the squash-vine were nearly circular, and slightly lobed, with a diameter of two feet and a half, and a superficial area of about seven hundred square inches. The leaf-stalks were hollow, two feet in length, and curiously marked with vertical striæ, alternately light and dark in color. The light lines were found to contain bundles of fibro-vascular tissue, while the dark ones were simple cellular tissue, containing chlorophyl.

The special functions of the leaf are to absorb carbonic acid from the atmosphere, and, by a process of digestion, form from its carbon and the elements of water, the soluble starch and sugar out of which the tissues of the plant are constructed; to exhale the surplus water of the crude sap, and thus aid in its ascension from the soil and the roots; to exhale the oxygen set free in the process of digestion, and thus to purify the air for the respiration of animals; and, finally, to exhale, at night especially, the surplus carbonic acid liberated within the plant in the process of vegetable respiration, which appears to be as necessary and constant as that of animals. It seems also most probable that the albuminoids, or protoplasmic substances, are first produced in the leaf, and thence transferred to the various localities, where they are needed in the process of growth.

To facilitate and control the absorption and exhalation of gases and aqueous vapor, leaves are furnished with breathing-pores, or stomates, which open under the stimulus of light and moisture, and close in darkness, or when scantily supplied with water. These stomates are about twice as numerous on the under as on the upper side of the squash-leaf, and the total number is about one hundred and fifty thousand to the square inch, or more than one hundred millions on each large leaf. One leaf of the great water-lily, *Victoria regia*, nine feet in diameter, contains about twenty-four hundred millions of stomates on its upper side, and none on its under surface, where they would be useless.

During the past year much has been written and said about carnivorous plants, which catch great numbers of insects for the apparent purpose of feeding upon them. When a fly alights on the leaf of a *Dionæa*, the two halves close upon it and hold it fast until consumed, when they open for another. The leaf of a species of *Drosera*, in New Jersey, is said to have the power of moving towards an insect, fastened within half an inch of it, and feeding upon it. The pitcher-shaped leaves of *Sarracenia variolaris* not only seem to possess the power of enticing insects to climb from the ground to the inside of their pitchers, by secreting a vertical line of honey on the outside, and also a line around the edge of the cup, but they prevent their escape by an ingenious arrangement of hairs, which continually force them downward as they attempt to fly out. When they thus reach the bottom of their prison, they come in contact with a fluid which first paralyzes them, and then hastens their decay and absorption.

Not less wonderful are the instinctive movements by which climbing plants seek for, and attach themselves to, a support. Twining vines, like the hop, the bean, and the morning-glory, exhibit a revolving movement of their extremities, until they come in contact with some object around which to coil. Each species has its own peculiar direction, from which most of them never vary. A few, like the hop, wind from the right upward towards the left, moving like the hands of a watch, but most, like the bean, move in an opposite direction. The squash, however, is not a twining plant, but climbs by means of tendrils. Nevertheless, the tip of a growing vine revolves continually from left over to right, in evident search for a support.

Mr. J. J. H. Gregory informs us, that if a shingle be set into the ground near the tip of a growing squash-vine, it will, in a day or two, be seen turning towards it; and that, if the shingle be removed to the opposite side, the direction of the vine will again be changed. He also states that he has observed a squash-vine, after running along on the ground ten or twelve feet, and then passing under the branches of a tree which were four feet above it, to stop and turn upward towards he branches. After growing in this direction till it could no longer sustain itself, the vine fell to the ground; but instead of proceeding horizontally, it again rose into the air, again to fail. A third effort was made before the plant was willing to give up and trail humbly on the earth.

The end of the vine under observation was constantly elevated to the sash-bars and glass above it, sometimes to the height of two feet, and as it increased in length, was pushed along against them. The extent and velocity of the terminal motion were doubtless greatest in August, when growth was most rapid. The record, however, was made in November. The time occupied in each revolution was variable, and the long diameter of the ellipse described, which was horizontal, measured about two inches.

The *tendrils* of the squash-vine were produced at the nodes, and the main stalk was hollow and divided into several branches at a point three or four inches distant from the vine.

These branches spread out in various directions, and attained a length of six or eight inches. Each branch gradually straightened out from the coil in which it first appeared, and increased in length. When about two-thirds developed, it began to revolve, so that its hooked tip described an ellipse several inches in diameter. Its revolution was made by a series of bendings, in such a way as not to twist itself. The tendrils moved in the same direction with the tip of the vine, but somewhat irregularly both as to time and to the figure described. During the day, the ellipse was broad, and at night, long and narrow. Usually, the motion was scarcely perceptible to the eye, but sometimes it moved two inches in five minutes. The average time of revolution in November was about three hours. If touched by the finger on the sensitive or inner side, the tendril bent towards the place where the finger was, and, not finding it, straightened itself again. If, however, it came in contact with any object to which it could cling, it bent at the point of contact, and the concave curvature extended along the inside of the branch, until the extremity was wound closely around the support. Other branches would, also, fasten to the same object, if possible. The tendril, thus attached, increased in size and firmness, and soon coiled upon itself in a double reversed spiral, so as to exert a strain on the support. All the branches having done this, they pull together and must fail together, if at all.

Another most obvious benefit derived from this double spiral, is the elasticity of the fastening, which greatly diminishes the danger of rupture by violence. If the tendrils of the squash failed in finding a support, the branches then coiled upon themselves, and the main stalk often turned back along the vine.

The habits of climbing plants have been studied by Mr. Charles Darwin and others; but this field for research is by no means exhausted.

The tendrils of the grape vine are not very sensitive, but fasten themselves very firmly to a suitable support. The tendrils of the *Cobœa scandens* are long, branching, and tipped with woody claws. They are extensions of the petiole of a compound leaf, revolve actively, and attach themselves in a most marvellous manner. When a revolving branch has found a support, it contracts so as to bring its extremities in contact with it. The other branches seek the same object, and, as they are sensitive on all sides, they fail in many cases to secure a firm attachment with their claws. They therefore detach themselves from their support, one at a time in succession, twist so as to bring their claws into the proper direction, and then again make fast.

It is well known that most plants grow toward the strongest light; but climbing plants are sometimes exceptions. English ivy turns its young shoots away from the light in order that they may come in contact with dark objects,—such as rocks and trunks of trees,—to which they then attach themselves by short roots. The tendrils of the Virginia creeper, or woodbine, are among the most wonderful. They grow away from the light, and send their branches into crevices of old bark and rocks. Sometimes such tendrils are said by Mr. Darwin to actually show a power of choosing one place of attachment in preference to another, by penetrating a cavity and then withdrawing to seek a more satisfactory one. As soon as the tendrils of the creeper find a support, the branches spread out their tips and press them against it. Little pads of hard cellular tissue are now developed at the points of contact, and the tendril coils on itself and becomes very tough and woody. At the end of the first season it dies, but remains firmly fixed to its support for many years. Mr. Darwin found one, which, though ten years old, was not detached by a weight of ten pounds from the wall to which it had adhered.

The chemical constitution of the squash-vine under observation has not yet been determined; but its anatomical structure, in all its parts, may be readily understood by an examination of the figures appended to this paper, which are accompanied by detailed explanations. The vine, the petioles, the flower-stalks, the tendrils and the fruits were hollow, so that about thirty per cent. of the apparent size was simply air. The greater proportion of the remainder was water, so that less than ten per cent. of the entire volume was solid, dry material. The large, yellow flowers were arranged in regular succession, one at each node. A female flower was usually succeeded by four males, so that on such a vine a squash would be produced at every fifth node, if every one should set, which, however, never happens. The impregnation of the ovules within the ovary of the female flower requires the deposition of pollen-grains from the anther-cells of the male flower upon the stigma of the former under favorable circumstances. The stigmatic surface must be in a proper condition to retain and develop the pollen, which must be in a perfect state. Bright, warm weather will doubtless aid in the process, though many observations are still needed concerning this subject. The pollen-grains of the squash are large and rough, and of a spherical form, and consist of an outer and inner coating of membrane filled with a protoplasmic fluid. In the outer coating is a minute orifice, through which, when moistened by the saccharine secretion of the stigma, the inner coating protrudes as a microscopic, structureless tube, which pushes its way through the tissues of the style and ovary until it reaches the embryo-sac of an ovule, which may then become a perfect seed. This contact of the pollen-tubes with the ovules is essential to the setting of every squash. The transfer of the pollen-grains to the stigmas is usually accomplished by insects which fly from flower to flower in pursuit of food. It may, also, be done artificially, and there is reason to believe that the crop of squashes, melons and cucumbers might often be largely increased by attention to this matter in out-door cultivation. When grown under glass, fertilization must always be effected by artificial means.

The pistillate, or female flower, on the twenty-first node of the growing vine, was artificially impregnated with pollen from a staminate, or male flower, on the first of August. The young squash immediately began to enlarge, and, on the fifteenth of the same month, measured twenty-two inches in circumference; on the sixteenth, twenty-four inches, and on the seventeenth, twenty-seven. Though the rind of the young fruit was very soft, it was now determined to confine it in such a way as to test its expansive power. In doing this, great care was taken to preserve the health and soundness of every part of the squash, and to expose at least one-half of its surface to the air and the light. The apparatus for testing its growing force consisted of a frame, or bed, of seven inch boards, one foot long. These were arranged in a radial manner, like the spokes of the lower half of a wheel, their inner edges being turned toward the central axis. These pieces were held firmly in place by two end-boards, twelve inches square, to the lower half of which they were secured by nails and iron rods. A hemi-ellipsoidal cavity, about five inches deep in the centre and eight inches long, was cut from the inner edges of the seven boards, and in this the squash was carefully deposited, the stem and vine being carefully protected by blocks of wood from injury by compression. Over the squash was placed a semi-cylindrical harness, or basket of strap-iron, firmly rivetted together. The meshes between the bands, which crossed each other at right-angles, were about one inch and a half square. The harness was twelve inches long and the same in width, so that when placed over the squash, it just filled the space between the end-boards. Upon the top of the harness, and parallel with the axis of the cylinder and the squash, was fastened a bar of iron with a knife-edge to serve as the fulcrum of a lever to support the weights by which the expansive force was to be measured. At first, an iron bar, one inch square, was used for a lever, then a larger bar of steel, then a lever of chestnut plank, then one of seasoned white oak plank, and, finally, one of chestnut, five by six inches square, and nine feet long; but even this required to be strengthened by a plate of iron four inches wide by half an inch thick and five feet in length. The fulcrum for the lever was also renewed from time to time as the weight was increased.

The following table shows the weight of iron lifted by the squash in the course of its development:—

August 21,	60	pounds.
" 22,	69	"
" 23,	91	"
" 24,	162	"
" 25,	225	"
" 26,	277	"
" 27,	356	"
" 31,	500	"

September 11,	. . . . .	1,100 pounds.
" 13,	. . . . .	1,200 "
" 14,	. . . . .	1,300 "
" 15,	. . . . .	1,400 "
" 27,	. . . . .	1,700 "
" 30,	. . . . .	2,015 "
October 3,	. . . . .	2,115 "
" 12,	. . . . .	2,500 "
" 18,	. . . . .	3,120 "
" 24,	. . . . .	4,120 "
" 31,	. . . . .	5,000 "

The last weight was not clearly raised, though it was carried ten days, on account of the failure of the harness irons, which bent at the corners under the enormous pressure of two and a half tons, and consequently broke through the rind of the squash. It was not feasible to remove the harness and substitute for it a stouter one, on account of its being imbedded in the substance of the squash, which grew up through the meshes of the harness, forming protuberances an inch and a half high and overlying the iron bands. When, on the seventh of November, the harness was removed in order to take a plaster cast of the squash, it was necessary to cut the straps with a cold-chisel, sometimes into several pieces, and draw them out endways.

The growing squash adapted itself to whatever space it could find as readily as if it had been a mass of caoutchouc; nor did it ever show the slightest tendency to crack, except in the epidermis. This would often open in minute seams, from which a turbid mucilaginous fluid exuded. In the morning drops of this would frequently bedew the protuberances like drops of perspiration. In the sunshine these dried up and fell off as minute globules, resembling gum Arabic.

The lifting power was greatest after midnight, when the growth of the vine and the exhalation from the foliage was least.

The material out of which the squash was formed was elaborated in the leaves during the day-time, and transferred through the vine to the stem. Through this it was imbibed by the living, growing cells of the squash, which were constantly multiplying by subdivision until their number was many billions, notwithstanding the enormous pressure under which they were forced to develop. This growth was possible only because life is a molecular force and exerted its almost irresistible power over an immense surface of cell membrane.

Scarcely less astonishing than the mechanical force exhibited was the ability of the tissues of the squash to resist chemical changes and the attacks of mould, when the rind was injured by bruises or cuts. Whenever fresh-growing cells were exposed to the action of the air, they immediately began to form a regular periderm of cork, precisely similar in appearance and structure to that produced upon the cork-oak, the elm, and other trees.

The form of the squash can hardly be described, but may be seen in the drawings which show the upper and under sides. The weight was forty-seven pounds and a quarter, and when opened the rind was found to be about three inches thick and unusually hard and compact. The internal cavity corresponded in general form to the exterior, but was very small, and nearly filled with fibrous tissue and plump and apparently perfect seeds in about the normal number. A squash of the same variety, grown in the field by Messrs. Russell Brothers, in North Hadley, weighed one hundred and twenty-three pounds. Its form was ovoid, but flattened as if by its own weight, and the cavity within had a capacity of about sixteen quarts.

Two vines having been started together in our experimental bed, it was decided to apply a mercurial gauge (such as will be described in another place) to the neck of one cut off at the ground, when the vine was about eight weeks old and had a length of twelve feet. The result was quite surprising, greatly surpassing anything heretofore recorded, so far as we are aware, concerning the pressure exerted by the sap of an herbaceous plant, the maximum force with which the root of the squash exuded the water absorbed by it being equal to a column of water 48.51 feet in height. The gauge was applied about noon, August 27th. At 2 P. M., August 28th, the temperature of the pit being 86° Fahrenheit, the pressure on the gauge equalled 31.70 feet of water.

PHENOMENA OF PLANT-LIFE. 21

At 4, p. m., Aug. 28, it was 29.47 feet, Temp. 75° Fahr.
" 9, "   "   28,   "   25.78   "   "   63°   "
" 7, a. m., "   29,   "   32.30   "   "   63°   "
" 2, p. m., "   29,   "   42.59   "   "   85°   "
" 9, "   "   29,   "   48.51   "   "   65°   "
" 8, a. m., "   30,   "   39.33   "   "   70°   "
" 12, m.   "   30,   "   35.25   "   "   84°   "
" 7, a. m., "   31,   "   27.88   "   "   67°   "
" 8, " Sept. 1,   "   00.00   "   "   00°   "

[For illustrations relating to the squash, see figures 1–16.]

Gauges were also attached to the stumps of large plants of Indian corn, tobacco, and the dahlia. The results were not specially different from what has been previously observed by Hofmeister and others. The flow continued but a very few days, and the pressure varied from eight to twenty-five feet of water. The pressure in all these cases seems to be caused by the activity of the absorbent tissues of the root; and its cessation results, doubtless, from the stagnation of the sap in the gorged cells and vessels, and the consequent decay of the root-hairs and fibres.

The frequent displacement of flagging-stones, and the damage often done to brick and concrete pavements and stone walls by the roots of shade trees, considered in connection with the wonderful expansive power exhibited by the squash in harness, made it evident that growing roots of firm wood must be capable of exerting, under suitable conditions, a tremendous mechanical force. Upon searching the fields for examples of trees standing upon naked rocks, or ridges covered with only a shallow soil, many interesting specimens were readily discovered to demonstrate this fact.

In South Hadley, Mass., a sugar maple was found which had grown upon a horizontal bed of red sandstone. The tree stood upon the naked rock, over which its roots extended a few feet in three directions into the soil. One root had pushed its way under a slab of rock which measured more than twenty-four cubic feet, and must have weighed about two tons. In the course of twenty years or more, this root had developed to such a size as to raise the slab entirely from the bed-rock and from the earth, and so that it rested wholly upon the wood. Upon examining the tree, it was evident that as it stood upon the horizontal roots which rested on solid rock and had a diameter of nearly a foot; and as they had grown by the deposition of an annual layer of wood entirely around them; and as the heart, now several inches from the rock, must once have rested on it; and as the rock could not have been depressed,—therefore, the tree had been lifted every year by the growing wood of the outside layer.

Another tree of paper birch having been found growing in a similar manner, one of the horizontal roots was sawed through, and the centre of the heart was seen to have been elevated seven inches since the tree was a seedling.

Mr. William F. Flint, a student in the Agricultural College of New Hampshire, has rendered valuable assistance in finding specimens of trees which illustrate this principle in an admirable manner. Drawings of two such examples selected from a large number furnished by him are appended to this paper. (Figs. 18, 19.)

Now it is clearly demonstrated that the power of vegetable growth can lift a tree, and that it must do so, whenever the bed of the roots cannot be depressed. It is evident also that old trees on a clay hard-pan or any other unyielding subsoil must be thrown up by the process of growth. Every person is familiar with the fact that large trees usually have the appearance of having been thus raised, and their roots are often bare for a considerable distance around the trunk.

This lifting of the tree from its bed would seem to be advantageous to it by tightening the roots so as to hold it firmly in place, notwithstanding the possible elongation of their woody fibre by the tremendous strains to which they are subjected during violent storms. This method of securing the tree in place would be still further improved by the constant enlargement of the roots by the annual deposition of a layer of wood, and the consequent filling of any spaces formed in the soil by the movements of the roots, caused by the swaying of the tree in the wind.

This slight annual elevation of trees by the increase in diameter of their horizontal roots furnishes an explanation for the differences of opinion in regard to the question whether a given point on the trunk of a tree is raised in the process of its growth. While it has been demonstrated by Prof. Asa Gray that two points in a vertical line on the trunk of a tree will not separate as it enlarges, it seems equally clear that both of them may be quite perceptibly elevated in the course of time.

It has been stated on good authority that, at Walton Hall, in England, a mill-stone was to be seen, in 1863, in the centre of which was growing a filbert tree, which had completely filled the hole in the stone, and actually raised it from the ground. The tree was said to have been produced from a nut, which was known to have germinated in 1812. The above story has been declared false, because, as asserted, the tree could not have exerted any lifting power upon the stone. It is, however, not difficult to see that it may be true, and is even probable.

Yet it should be remembered that the amount of elevation, in any case where it occurs from the increase in the size of horizontal roots, must depend upon the firmness of the material on which they rest, and can never exceed one-half the diameter of the largest roots. When, therefore, a writer, as has happened, asserts that, during a visit to Washington Irving at Sunnyside, he carved his name upon the bark of a tree beneath which he was sitting in conversation with the illustrious author, and that many years after he went to the place, and with much difficulty discovered the identical inscription, high up among the branches, far above his reach, it is altogether probable that his feelings were too many and too exalted for the ordinary use of his intellectual faculties.

Since the publication of the paper on the "Circulation of the Sap in Plants," in the last volume of the Agriculture of Massachusetts, a course of lectures on the "Physiology of the Circulation in Plants, in the Lower Animals, and in Man," by Dr. J. Bell Pettigrew, has been published by Macmillan & Co., of London. The hypotheses adopted by this author are quite extraordinary, and evidently announced without the slightest attempt at demonstration, although he has invented a new method of accounting for the phenomena of the motions of the sap. Thus he says, "In trees the sap flows steadily upward in spring, and steadily downward in autumn." Also, "Much more sap is taken up than is given off in spring, in order to administer to the growth of the plant. In autumn, when the period of growth is over, this process is reversed, more sap being given off by the roots than is taken up by them." Now, this is pure assumption, there being no proof that the sap of trees escapes from the roots in autumn.* In fact, it appears that the wood of trees contains as much sap in winter, when at rest, as in the period of most active growth.

Again, Dr. Pettigrew remarks : " It is difficult to understand how excess of moisture in the ground can be drawn up into the plant and exhaled by the leaves at one period, and excess of moisture in the atmosphere seized by the plant, and discharged by the roots at another. The explanation, however, is obvious, if we call to our aid the forces of endosmose and exosmose. The tree is always full of tenacious, dense saps, and it is a matter of indifference whether a thinner watery fluid be presented to its roots or its leaves ; if the thinner fluid be presented to its roots, then the endosmotic or principal current sets rapidly in an upward direction ; if, on the other hand, the thinner fluid be presented to its leaves, the endosmotic or principal current sets rapidly in a downward direction."

This explanation is not only false, but superfluous, since no such circulation can be shown to exist, but is an excellent sample of the common mode of dealing with this obscure subject. Instead of seeking to discover the exact facts concerning the composition and movements of the sap in all parts of the plant, a display of book-knowledge is made by quoting from numerous writers of some repute, such statements as seem to corroborate the hypotheses of the author. The assumed phenomena of the circulation are then accounted for in an apparently scientific manner by ingenious allusions to osmose, capillarity, and other physical forces, the surprising possibilities of which are duly recounted.

Dr. Pettigrew further observes, that "Herbert Spencer believes that the upward and downward circulation of crude and elaborated saps takes place in a single system of vessels or vertical tubes." To explain this extraordinary assumption, Mr. Spencer states that "the vessels of the branches terminate in club-shaped expansions in the leaves, which expansions act as absorbent organs, and may be compared to the spongioles of the root. If, therefore, the spongioles of the root send up the crude sap, it is not difficult to understand how these spongioles of the leaf send down the elaborated sap, one channel sufficing for the transit of both." This hypothesis concerning the circulation of sap is accepted only by its inventor, and is directly opposed to most of the facts of plant-growth.

Finally, Dr. Pettigrew has conceived a system of syphons by the aid of which he is able to account to his entire satisfaction for all he knows concerning the circulation of sap. He says : "The vessels which convey the sap, as is well known, are arranged in more or less parallel vertical lines. If the vessels are united to each other by a capillary plexus, or, what is equivalent thereto, in the leaves and roots, they are at once, as has been shown, converted into syphon-tubes, one set bending upon itself in the leaves, the other set bending upon itself in the roots. As, however, a certain portion of the syphon-tubes, which bend upon themselves in the roots, are porous and virtually open towards the leaves; while a certain portion of the syphon-tubes, which bend upon themselves in the leaves, are porous and virtually open toward the roots,—it follows that the contents of the syphon-tubes may be made to move by an increase or decrease of moisture, heat, etc., either from above or from below. In spring, the vessels may be said to consist of one set, because at this period the leaves and the connecting plexuses which they contain do not exist. All the vessels at this period may, therefore, be regarded as carrying sap in an upward direction to form shoots, buds and leaves, part of the sap escaping laterally, because of the porosity of the vessels. In summer, when the leaves are fully formed, the connecting links are supplied by the capillary vascular expansions formed in them,—the tubes are in fact converted into syphons. As both extremities of the syphons are full of sap in spring and early summer, an upward and a downward current is immediately established. When the downward current has nourished the plant and stored up its starched granules for the ensuing spring, the leaves fall, the syphon structure and action is interrupted, and all the tubes (they are a second time single tubes) convey moisture from above downward, as happens in autumn. As the vascular expansions or networks are found also in the stems of plants, it may be taken for granted that certain of the tubes are united in spring, the upward rush of sap being followed by a slight downward current, as happens in endosmose and exosmose. As, moreover, the spongioles of the roots and the leaves are analogous structures, and certain tubes are united in the roots, the downward current in autumn is accompanied by a slight upward current. This accounts for the fact that at all periods of the year, the upward, downward and transverse currents exist; the upward and downward currents being most vigorous in spring and autumn, and scarcely perceptible in winter. Furthermore, as some of the vascular expansions in the leaves are free to absorb moisture, etc., in the same way that the spongioles are, it follows that the general circulation may receive an impulse from the leaves or from the roots, or both together, the circulation going on in a continuous current in certain vessels."

This original effort of the learned lecturer on physiology, at Surgeons' Hall, in Edinburgh, published in 1874, to explain some of the most difficult problems of vegetable life by a mere hypothesis, which assumes that sap flows in the vessels; that there are spongioles in the leaves which absorb water; that the sap descends to the roots and escapes from them in autumn; and that an imaginary system of syphons does all these wonderful things, which have not been proved to occur at all, and which well-informed physiologists are almost unanimous in denying, reminds us of the adage that "a prophet is not without honor save in his own country." This is not the method of the Baconian philosophy.

In the observations which follow, we hope to add some new facts to the knowledge of the world concerning the phenomena of plant growth; but are painfully conscious of the need of much more investigation before a complete and correct theory of the circulation of sap can be stated. Exceptions have been taken to the use of the expression "circulation of sap"; but since there is an evident distinction between the crude and elaborated saps, both in their composition and their location in the plant, at least in the higher forms of vegetation, and since the circulation of blood is accepted as a proper term even when applied to animals without a heart, we prefer to retain it in our vocabulary.

In regard to the causes which induce the absorption of water and soluble substances by the roots of living plants, it seems unfortunate that so much has been claimed for osmose in this connection. Boussingault has recently shown that roots containing sugar do not exude it when growing in water, while leaves and fruits, when immersed in this fluid, readily absorb it by an osmotic process and part with their sugar: If the enormous absorption of water by the roots of birch trees, in spring, were accompanied by any corresponding exudation, it would appear easy to find it; but no one has yet detected it. It is not possible to account for the fact that when sap is rising most rapidly, none will flow from a wound in the bark, even when it will run a stream from the outer layer of wood, if the circulation in the trunk is caused by osmose. There is fresh cellular tissue in the liber, and some soluble material, but the bark remains comparatively dry till growth begins. After the cambium has become abundant, why should not all the crude sap press toward it and draw the elaborated material directly into the wood, instead of pushing its way against the force of gravity to the leaves, if osmose is so powerful an agent in the circulation? If this tendency to press into the bark were to exist, there would be a much greater flow from places that are girdled than is now observed; and probably the bark itself would be ruptured by the pressure exerted, which would often be equal to more than thirty pounds to the square inch.

One of the most surprising facts to be noticed in examining the wood of any tree with well-developed foliage, is the entire absence of anything like free or fluid water. A freshly-cut surface of the sap-wood is not even moist to the touch; and if a tube be inserted into the trunk of such a tree, it will frequently absorb water with great avidity. On the sixth of June last, a half-inch tube six feet in length was attached to a stopcock inserted into the trunk of an elm and the tube filled with water. The absorption was so rapid that the fluid disappeared in thirty minutes, and this was repeated several times the same day. Similar observations were made upon white oak, chestnut and buttonwood trees.

Now the absorption was not osmotic, since the rapidity of it was too great and there was no outward flow, but apparently the result of imbibition, or the affinity of the cellulose of the woody fibre for water. Is not this, then, the proper name for the force which carries up the crude sap?

The wood of growing trees when cut from near the surface, though apparently dry, contains nearly fifty per cent. of water; and in the young twigs, with a living pith, the proportion is even greater. A number of analyses have been made of specimens collected at different seasons during the past year, of which a tabular statement is appended.

There is good reason to believe that the sap in ordinary trees begins to move first in the buds, and that the first supply of water exhaled in the spring is derived from the sapwood. Branches of aspen and red maple, two feet in length, were cut on the twenty-sixth of March and placed in a warm room in an empty vase. The flower-buds developed without any other water than what they could abstract from the wood, so that on the fifth day the staminate catkins of the aspen were four inches long, and the pollen well developed. It is by no means uncommon to see large branches, which have been removed from apple trees early in the spring, covered with blossoms in a similar way while lying on the ground.

It is a well-established fact that the roots of most woody plants have not power at any season to force water to any considerable height when separated from their stems. Upon this point a large number of observations have been made, which will be described in another place.

The roots of all plants growing on ordinary soil develop most freely and absorb most abundantly when the earth is well drained and aerated. Thus we find that the crude sap imbibed by the root-hairs from the surface of the particles of the soil seems to be taken up in a dry state, that is, it appears to be absorbed molecule by molecule, no fluid water being visible, and carried in this form through all the cellulose membranes between the earth and the leaf, by which it is to be digested or exhaled. We do not say this is literally true, but it accords very nearly with what is constantly to be seen in some species of plants. The circulation of the sap in a poplar tree is very dry compared with that of the blood of any animal. Not a drop of moisture will ever flow from the wood of an aspen, so far as we have observed. Nevertheless, it grows very freely and starts very early in the season.

PHENOMENA OF PLANT-LIFE.

That living cellulose has a peculiar and very powerful affinity for water is evident from the experiments of De Vries, who discovered that when a shoot of an herbaceous plant with large leaves is cut, and the fresh surface allowed to come for a short time into contact with the air, it loses much of its absorbing power and the leaves wilt. If, however, the section be made under water, so that the living tissue is not exposed to the air, its power of imbibition remains unimpaired, and the leaves do not wilt.

It appears, therefore, that much of the crude sap passes through the membranes of the sap-wood or woody fibre or cellular tissue of plants in an apparently solid form, combined with the cellulose, just as the water in dry slacked lime, or a plaster cast is in a solid form. In all these cases it may be obtained as a liquid by distillation at a temperature of 212° Fahrenheit. The cause of the motion seems to be the removal of the water from the tissue at some point by exhalation, by chemical combination or by assimilation. Whenever any portion of the living cellulose has an insufficient amount of water to saturate its affinity, it imbibes an additional quantity and this process is continued from cell to cell downward, or backward to the roots and the earth.

The conducting power of the cellulose of sap-wood is very remarkable, as is seen in the fact that whenever a limb of an apple or peach tree breaks down under its burden of fruit, it very rarely wilts or fails to ripen its crop. Those who have compared the area of a section of the trunk of a large tree with the area of a section of its branches at any point above, must have noticed that the relative amount of sap-wood rapidly increases as we ascend toward the top, the young twigs and branches containing no other wood.

An elm in Amherst, famous for the beautiful symmetry of its form and known as the Ayres elm, was carefully measured by Prof. Graves and the senior class. The area of the sections of the branches twenty feet from the ground was more than twice as great as the area of a section of the trunk four feet from the earth, and the proportion of sap-wood was of course much greater.

An interesting experiment was undertaken in the Durfee Plant-house to determine how small a proportion of sap-wood could conduct the necessary supply of sap to the foliage of a growing tree, and also whether the bark alone could furnish the requisite water to prevent the leaves from wilting. A specimen of *Hibiscus splendens*, standing in the ground and having three stems from the same root, was selected for trial. The shrub was growing rapidly, and was prepared for the experiment as follows: Two of the stems were tied firmly to stakes, and the third left undisturbed. The first specimen had all the bark removed from one inch of the stem, and then the wood was cut away till there remained only a small piece of the outside layer of sap-wood, which was one inch long and seven-sixteenths of an inch in circumference. This exposed surface was immediately covered with grafting-wax, to protect the tissues from the action of the air. The amount of stem remaining was just one eighty-fourth of the original, which was about four inches around. The healthy leaf-surface was fully twenty-five hundred square inches, from both sides of which exhalation went on to some extent, making five thousand square inches of exhaling surface. The result was, that the foliage remained perfectly fresh and vigorous for ten days, until, on the tenth of November, the specimen was cut for the museum. (Figs. 20, 21.)

The other stem was used to determine whether by osmose, or in any other way, the crude sap could ascend in the bark and supply the leaves with water. All the wood and one-third of the bark were removed from a portion one-half inch in length, the exposed tissues protected by wax, and the branches so pruned as to leave only five hundred square inches of leaf-surface. The foliage all drooped in a single hour and never recovered. This experiment showed that the bark was altogether incompetent to furnish the requisite supply of crude sap to the parts above it, although it was thick and succulent, and much greater in quantity, when compared with the exhaling surface, than the piece of sap-wood which showed such marvellous conducting power. If osmose were the cause of the ascent of sap, it would seem that the abundant parenchyma of the bark, intimately united as it is with the wood by the medullary rays, must freely transmit the amount required in this case. But the leaves wilted and perished as quickly as if the entire stem had been severed.

Having thus demonstrated that crude sap ascends chiefly in the sap-wood of exogenous trees, let us now consider a few facts which appear to prove that there is a counter-movement of elaborated sap which is for the most part confined to the bark.

It is well known that if a narrow ring of bark be removed from the trunk of a tree between the leaves and the roots, then the deposition of wood ceases below the girdled place, though above it the growth for the season ensuing will be quite normal. This proves beyond dispute that the wood cannot convey that portion of the elaborated sap which is essential to growth, and that it can be conducted only by the tissues of the bark, or the imperfectly-developed tissues of the cambium between it and the perfectly-formed wood. Nevertheless, there is free communication in a transverse direction for the crude sap and for some of the elaborated substances between the wood and the bark, probably by means of the medullary rays which connect the two. Thus only can we account for the fact that the bark below a girdled place remains alive long after the deposition of wood ceases, and also for the circumstance that starch and sugar, which must originally come from the leaves, are found either accumulated in the cells of certain stems and roots, or existing in the sap which flows or is expressed from their tissues. If we shave off, little by little, the bark of a maple when the sap is flowing freely, we shall observe no exudation from any portion of the liber, even, but as soon as the whole of this is removed, the sap issues from every part of the surface.

Again; those who work with mill-logs tell us that in the spring the bark becomes soft and loose, precisely as if the tree were standing, at least in the case of some species. Sometimes logs and poles, cut for fences, will sprout and actually produce shoots with foliage, the sap of which must be derived wholly from the timber, and must, therefore, pass from the wood to the bark.

Mr. Wm. F. Flint has sent us a piece of a red maple slab, which he found on moist ground, under a pile of wood, and which had thrown out at the ends and sides a callous a quarter of an inch thick, precisely like an ordinary cutting of a grape vine. Here we have an instance of growth without either roots, buds, or leaves, all the material for which must have been derived from the stick itself. (Fig. 22.)

Similar to this in character is the curious circumstance, not very unfrequent, of old potatoes resolving themselves into several smaller ones, within the skin of the parent tuber, without any external appearance of vegetation. This is reported to have occurred in a vast number of tubers, in a quantity of potatoes on board a vessel in the Arctic ocean, where the low temperature probably exerted some influence in causing this peculiar mode of sprouting.

An excellent demonstration of the transverse diffusion of sap was obtained in some experiments performed to observe the result of protecting girdled places on trees from the effects of exposure. Healthy young trees, or large branches, of elm, chestnut, apple, grape, and white pine were drawn through glass tubes, two inches in diameter and two feet long, upon either end of which were fastened short pieces of rubber hose. These tubes were placed over girdled spots, from which the bark was removed on the thirtieth of May last, and the rubber securely fastened with iron wire to the tree. From all of these specimens a considerable quantity of sap escaped, apparently in the form of vapor, and was collected in the tube. There was no layer of wood formed, but the foliage of all except the pine was killed before autumn, apparently by the fermentation of the sap and its re-absorption into the wood. In the case of an elm root, treated in a similar manner, the bark was renewed, probably from the fact that the cambium was in a more advanced state than in the other instances. The root was dug up with care, twenty feet of it drawn through the tube, and then covered again with earth. (Fig. 23.)

With the view of determining some facts concerning the functions of the bark in connection with the circulation of sap and the growth of wood, many experiments have been undertaken at the College during the past two years, and some interesting results obtained.

In order to learn whether the annual layer of wood upon trees is developed from the outside of the old wood or from the inside of the bark, the following plan, suggested by the interesting experiments of Duhamel more than a century ago, was tried. Vigorous young trees of elm, glaucous willow, and chestnut were selected, which were from two to three inches in diameter. On the thirtieth of May, before any deposition of recently organized tissue was visible, but when the bark was easily separated from the wood, a horizontal incision was made with a sharp knife around each stem, and immediately above this four vertical incisions on the four quarters of the stem about three inches in length. The four strips of bark were then carefully detached from the wood at their lower ends, and a piece of tinned copper, one inch wide, and long enough to reach around the wood and overlap, was adjusted to the trunk. The bark was then replaced and covered tightly with cloth which had been dipped in melted grafting-wax. The trees grew through the season as usual, and after the fall of the leaves the bandages we e removed and the results observed.

In all cases the new wood was found to have been deposited from the bark and outside of the metallic band. Examination under the microscope showed that a thin layer of parenchyma, corresponding to the pith of the first year's wood and such as probably unites all the layers of wood in exogenous stems, was formed upon the metal, and outside of this the fibro-vascular tissue, while the medullary rays were as numerous as in the other portions of the layer of wood, and extended directly from the bark to the metal under it, whether examined in a transverse or a longitudinal section,—thus proving that the material did not flow down in an organized condition from above the band. The figures appended will render the entire experiment sufficiently intelligible. (Figs. 24–27.)

This quite satisfactory result demonstrates that the elaborated material formed in the leaves descends altogether outside of the wood, and that the inner bark is the most highly vitalized part of the trunk of a tree and the source of the new layers of wood and bark which are annually produced.

Much information has also been obtained in regard to the effects of ringing or girdling the trunks and branches of trees by the removal of a band of bark only, or of bark and sapwood from the entire circumference.

This has long been practised in new countries to kill the timber which the settler had not time to fell, but must destroy to obtain grain and other crops.

The Chinese are said to produce curious dwarf fruit-trees by ringing a fruit-bearing branch and placing over the spot a flower-pot with earth in which roots are developed, so that it may then be detached from the parent tree and cultivated independently. The Italians propagate the fig-tree in a similar manner, and this process may be made very useful in securing the certain growth of a sporting branch of any woody plant, or of the branches of species with spongy or pithy wood which will not root from cuttings. It is a well-known fact that the ringing of a branch of a vine or tree will tend to increase the size of the fruit the following season, because the branch is thereby gorged with elaborated material for which there is no outlet, and some persons habitually adopt this mode of improving their fruit.

In the town of Southborough, Mass., is an apple orchard of healthy trees, from twelve to sixteen inches in diameter, which were all girdled by the owner, Mr. Trowbridge Brigham, in the spring of 1870, for the purpose of inducing fruitfulness. The desired result is said to have been obtained, and the trees seem to have suffered no material injury, owing to the imperfect manner in which the operation was performed. At the time when the trees were in full blossom, a narrow belt of bark, usually less than an inch in width, was removed from the trunks, about two feet from the ground. This did not peel freely in all cases, and there were many crevices where it was retained. By means of these connecting links, the communication between the leaves and the root was imperfectly preserved, and during the season new wood and bark were developed upon these places. In addition to this, in many cases, the new wood from the upper side of the girdled spot was sufficiently abundant to reach across and form a connection with the living bark below.

Upon one of these trees was found a branch some four inches in diameter, which had been perfectly girdled in 1870, and, although no communication had existed between the bark of the branch and that of the trunk, it had grown every year till March, 1874, when it was cut. The buds upon it were poorly developed, but alive, and the ends of the branches were dead. It apparently could not have survived more than a year or two longer, and the reason was obvious upon mak-

PHENOMENA OF PLANT-LIFE.

ing a longitudinal section through the girdled part. The limb was nearly horizontal, and the ring of bark removed was only a few inches from the trunk. New layers had formed each year up to the denuded place, but the enlargement was more above this than below it. The material to form new wood and bark below came from the other parts of the tree, and yet, owing apparently to the poor circulation, was deficient in quantity. The crude sap with some materials from other portions of the tree ascended to the buds and leaves, and so an unhealthy growth was continued. An examination of the figure representing a section of this branch will explain the cause of its final failure. The wood through which the sap must ascend was gradually dying, and thus the channel of communication was constantly becoming more and more obstructed. On the whole, this method of treating orchards cannot be recommended for general use. (Fig. 28.)

In regard to the length of time during which a perfectly girdled tree may continue to live, we have obtained some facts worth recording.

In India, it is necessary to girdle the teak trees the year before cutting them, in order to have them die and lose a portion of their sap by evaporation, since otherwise the logs will not float down the rivers to market. Removing a ring of bark is not sufficient to accomplish this result, and it is necessary to cut through all the sap-wood so as to prevent the ascent of water to the leaves.

Mr. W. F. Flint has communicated an interesting account of a beech tree about eighteen inches in diameter, which grew in an open pasture in Richmond, New Hampshire. It was girdled for the express purpose of killing it, in 1866, by chopping a gash two or three inches wide and nearly as deep entirely around the trunk near the ground. The next year it sent up sprouts from below the girdle and formed a new layer over its entire surface. This was repeated in 1867, but in 1868 the bark and sprouts of the lower part died, and dead branches began to appear in the top. This process of decline continued, and in 1873 but one of the large branches put forth its leaves; and, finally, on the ninth year (1874) it died utterly. This remarkable tenacity of life is doubtless due to the close, fine texture of the timber, and the fact that such beeches in open land have an unusual amount of sap-wood, and are hence called white beeches.

A red maple, on the College farm, which was girdled in April, 1873, by cutting a channel in the sap-wood two inches wide and one inch deep, bled most profusely, but grew as usual through the season. No wood, however, was formed below the girdle, and the bark died and separated from the wood. The roots, nevertheless, remained alive, and the tree has borne its usual amount of foliage during the summer of 1874, and formed its buds for next year, and produced a new layer of wood above the girdle. Specimens have been collected for chemical and microscopic analyses of the roots and of the wood and bark above and below the girdle, in the hope that some light may be thrown upon the subject of sap circulation and the functions of the bark, whenever this work can be done.

On the third of June last, branches of the apple, pear, peach, crab-apple and grape were girdled by removing a ring of bark one inch long. They grew well and bore an abundance of fine fruit, as was expected.

On the fourth of June, small trees of red maple, elm, aspen; willow, linden, chestnut, white pine, black birch, butternut, and a large wild grape vine, were girdled by removing a ring of bark two inches in length.

On the twelfth of June, trees of ash, bass, beech, black birch, yellow birch, white birch, alder, black oak, chestnut, sugar maple, hornbeam and ironwood were girdled in like manner; and on the twenty-third of June, specimens of white oak, red oak, black birch, yellow birch, white birch, red maple, sugar maple, ash, bass, aspen, witch-hazel, white pine, cornel, chestnut, hickory, beech, ironwood, hornbeam, apple and choke-cherry. July twenty-first, we girdled specimens of wild grape, cornel, red maple, chestnut, black birch, white birch, white pine, bitternut, white oak and black oak.

On the twenty-eighth of August, the bark of the following species was found to adhere to the wood, viz.: red maple, yellow birch, wild thorn, hornbeam, beech, witch-hazel, bird-cherry, white oak, red oak, elder and elm; while the bark of the following species was readily separated from the wood, viz.: hemlock, white pine, alder, shadbush, white birch, black birch, chestnut, cornel, ash, ironwood, apple and aspen.

All the trees thus girdled grew through the season as usual, but none of them formed wood below the girdle, except the grape and the red maple. The former, being a branch of a large vine, with foliage both above and below the girdle, formed new wood on both sides of it, and finally, the two callouses were united and communication restored across it. (Figs. 29–30.)

The red maple, girdled June 23d, formed wood only on the upper side, but the specimen girdled July 21st, formed a new layer of wood and bark upon the denuded surface. This was doubtless owing to the fact that a portion of the cambium was left on the wood sufficient to conduct the elaborated sap and form new tissues out of it. This tree, like the others, grew in the woods, where it was shaded from the direct rays of the sun. The new bark was of a reddish brown color and very smooth, and consisted of a thin layer of periderm or cork, with parenchyma and bast. A drawing of its microscopic structure together with one of the old bark on the same tree has been prepared. (Figs. 31–34.)

There is a popular notion that the bark of an apple tree, removed on the longest day of the year, will be renewed, and it is well known that occasionally such renewal of the bark of various species does occur. This may happen whenever there is deposited upon the old wood enough of the new layer to conduct downward the elaborated sap, and to develop from the living parenchyma of the forming medullary rays a protecting layer of periderm.

It is not uncommon for the bark of the half-hardy weeping-willow to be started by freezing and thawing from the wood. When this is the case, there sometimes forms a new layer of wood upon the detached bark, which is disconnected from the wood of the parent trunk. There is also sometimes formed a new layer of wood and periderm on the old wood under the shelter of the old bark, and roots often descend from the healthy portion of the trunk several feet beneath the loose bark to the ground, and as soon as they penetrate it enlarge rapidly. All these phenomena are readily explained by supposing that the liber, or inner bark, of the tree is torn asunder, a portion sometimes remaining attached to the wood sufficient to conduct the elaborated sap and so form a new layer of wood with a layer of bark. The roots are developed from the uninjured portion under the protection of the old bark, and in their nature are precisely like roots from cuttings. An interesting specimen from the grounds of Mr. Charles S. Smith, of Amherst, is exhibited in figure 35.

The rupture of the medullary rays and separation of the bark from the wood by the combined action of frost and sunshine is not uncommon in the apple and other cultivated trees. If a severe frost separates the water from the wood as ice, and it then thaws and freezes again before it can be reabsorbed, it will be likely to burst the bark or tissues in which it is accumulated. This usually results in one or more cracks through the bark on the southerly side of the tree, from which there is, in the case of the apple tree, commonly a slight flow of crude sap in the following April or May. The outside of the bark is blackened, and the detached portions die.

In the spring of 1874, a vertical crack three feet long was noticed in the south side of a vigorous young Gravenstein apple tree in Amherst, the trunk of which was about three inches in diameter. Upon examination, it was found that the bark had not been separated from the thick layer of wood formed the previous year, but that this outside layer was entirely detached from the wood beneath. The bark, being supplied with sap ascending through this layer, remained sound, and, the crack having been filled with wax, the tree grew equally well with others in its vicinity which had sustained no injury. The new growth on the sides of the crack being covered only with a thin, soft periderm, will, doubtless, readily unite, and there will soon remain no trace of the rupture. The separated layers of wood, however, will never be reunited, though the inner ones may conduct sap, until converted into the nearly impervious heartwood which occupies the central portion of every trunk after it attains to any considerable size.

At what age, if ever, the inner wood of exogens loses all power of conveying sap, and whether the sound heart of an old tree which has never been exposed to the influences of the

PHENOMENA OF PLANT-LIFE.

atmosphere still retains life, are questions which have not been definitively answered. It is not easy to say wherein the vitality of any perfectly formed tissue, whether of the wood or bark, consists, since their cells have no power of enlargement or multiplication, though the thickening of the cell-walls by the deposition of substances within the cells and the striking changes in color, seem to indicate the presence of a feeble life. The functions of the wood seem to be mainly such as may be performed by dead material. The cellulose which has never been exposed to the air may retain its peculiar affinity for water, which is evidently much greater before than after drying. The cells may serve as reservoirs of starch and other substances which may afterwards be imbibed by the living, growing or ripening tissues. The pith, which is alive in young branches so long as leaves are borne upon their wood, dies apparently with them. If growth is a characteristic feature of living tissue, our trees may with some reason be considered annuals, since all their growth proceeds normally from their winter buds and completely envelops every portion of the tissues of the roots, stems and branches previously formed, thus excluding them from the weather and preventing their decay, while using them for a support and a magazine of supplies. However this may be, it is certain that the vitality of trees is concentrated in a remarkable manner upon the surface and the extremities of their roots and branches.

Among the observations made during the past season, not the least interesting were those relating to the natural grafting which is frequently to be seen in the forests, and which is particularly noticeable among roots. The almost incredible manner in which the living surface of the inner bark of woody stems can transform the same elaborated sap into different species of wood and bark, was alluded to last year, and the case mentioned of a possible compound tree, containing a plum root and base, on which grew a stem of apricot, surmounted by a stem of blood peach with red wood, and that by a stem of white peach, and the whole by a stem and branches of almond. Thus, each kind of wood and bark would be perfectly developed from the same material, just as on the same cow's milk may be fed a child, a calf, a colt, a black pig, a white pig and a lamb. The specific life of each, and not its food, determines its form, size and character.

To show still more impressively the peculiar powers of the wood and bark to conduct the crude and elaborated saps in either direction, and to act either as roots or branches, as circumstances require, we will describe an experiment performed by a French gardener, M. Carillet, at Vincennes, in 1866 and 1867. He selected two dwarf pear trees, grafted on quince roots, which were from four to five feet high. One of them was carefully dug up in April, 1866, and fastened in an inverted position above the other. The leading shoots of the two trees were now flattened on one side with a knife, and the two surfaces firmly bound together in the usual manner of splice grafting. The two shoots grew together, and, in the course of the summer following, a few leaves appeared on the main stem of the inverted pear tree, and also on the main branches of the quince roots, which were entirely in the air some eight or ten feet from the ground. The next spring, scions from four varieties of pear were set upon the four main branches of the quince roots, two of which lived and grew several inches. Meanwhile, the inverted pear tree bore two pears. Here, then, was a composite tree, consisting, first, of a root of quince, then a pear tree, upon this an inverted pear tree, which had branches consisting of inverted quince roots, and these were surmounted by pear-shoots of two unlike kinds. Upon such a specimen it would be very difficult to comprehend the working of the imaginary syphons of Dr. Pettigrew, already described.

In order to illustrate the fact that the return of the elaborated sap was not the result of the force of gravity, a pendant branch of weeping-willow was girdled last June. The enlargement was on the lower side of the girdled place, showing that the flow of the material formed in the leaves was constantly towards the roots. (Fig. 36.)

To learn whether sap would flow from the bark on the upper side of a girdled place, a stem of white willow, an inch in diameter and ten feet high, was selected, and a ring of bark, one inch long, removed. The girdled place was then wrapped in oiled paper, so as effectually to exclude the air and the light. On the fifteenth of October, one month after girdling, the paper was taken off and the specimen examined. The wood appeared dead and brown, and was covered with a mucilaginous fluid which appeared to have come from above. There was no sign of growth below the girdle, but above it the stem was decidedly enlarged, and a callous had descended a quarter of an inch and developed upon itself a bud, as if about to strike out for air and light. No bleeding from the bark was observed in any case worthy of mention, the nearest approach to it being in the flow of turpentine from the bark and sap-wood of the the white pine.

Among the specimens of natural grafting obtained during the past year, perhaps the most remarkable was a fine bunch of mistletoe, growing as a parasite upon a branch of oak. This was kindly procured for the College museum by Prof. J. W. Mallet, LL.D., of the University of Virginia. The shrub is an evergreen, and its roots penetrate the bark and sap-wood of the tree on which it feeds, appropriating the crude sap and forming a wood of a totally different sort from that of its support, and having an ash peculiar to itself. In fact, the several species on which it is produced seem to serve merely as so many different soils on which it can thrive. As the oak-branch was dead beyond the mistletoe, it would seem to have been injured by the abstraction of its sap and its exhalation from the foliage of the parasite. The singular mode in which the union is formed will be understood by an examination of figure 37.

A specimen of red maple was brought to the College by Mr. Austin Eastman, of Amherst, which exhibited a single trunk with one heart, formed by the natural union of two shoots, which were nearly three feet apart, and were united about six feet from the ground. The main trunk was eight inches in diameter.

Another specimen, found in Pelham, shows two white pine trunks, joined like the Siamese twins, at about four feet from the ground. This, when sawed open vertically, showed how the union had been effected. A branch of one had lodged in the angle made by a branch of the other with its parent trunk. As the tree grew, they were fastened together, and, under the pressure thus caused, the bark was flattened until it almost disappeared, and soon the new wood formed over the scar and made the grafting complete. The structure will be understood by examining figure 38.

But the grafting of roots is still more common and curious. They seem to cohere without the least difficulty, especially those of the white pine, which is doubtless owing to the softness of the bark and young wood, and the fact that they grow so nearly at the same level in the earth. A specimen from the vicinity of the College, exhibiting a large number of grafts, is represented in figure 39.

A branch of gray birch, which has united with its own trunk by an attachment formed in the angle of another branch above it, is shown in figure 40.

The rootward flow of elaborated sap is well illustrated in a specimen of aspen in the College museum, around which is twined a vine of bittersweet. (Figs. 41–42.)

From the observations above made, it will be seen that there is no difficulty in accounting for the curious fact which has long been regarded as a great mystery, that the stumps of fir trees, which do not sprout, have been known to continue forming new layers of wood and bark for a great number of years. Dutrochet mentions the case of a stump of the silver fir which thus grew from 1743 till 1836, when it was still alive, having formed, since the tree was felled, ninety-two thin layers of wood. The roots of the living stump were doubtless grafted to the roots of some healthy tree or trees in its vicinity, and their elaborated sap was attracted into the sound bark and supplied the necessary material for the development of new tissues under the influence of its vital force. The outer layer of the roots of the stump was thus renewed annually, and so they retained their power of absorption; but since the top of the stump, becoming dry and having no foliage, could not exhale moisture, the crude sap of its roots ascended into the neighboring tree or trees to which they were united. Thus a sort of circulation was maintained sufficient to explain the phenomena observed.

Another peculiarity often to be seen in the stems and branches of trees and shrubs, as in the pear, the apple, the hemlock, and the lilac, is the spiral growth or twisting of the wood and bark, which is sometimes visible during the life of such specimens, and always when the bark is removed and the timber seasoned. Some have endeavored to account for this phenomenon by referring it to the effect of the wind, but it is frequently seen on trees which grow in sheltered situations. The timber of *Pinus longifolia*, a valuable tree of Northern India, is often rendered worthless by this habit of growth, and while such trees are more numerous in some regions than in others, they are found irregularly scattered among those which do not exhibit this abnormal structure. (Fig. 43.)

The surprising phenomena of pressure and suction exerted upon mercurial gauges attached to the trunks and roots of such trees as bleed or flow from wounds in the spring, which were described in the paper presented to the Board last year, gave abundant encouragement for further investigation. Accordingly, numerous experiments have been undertaken and some thousands of observations recorded, which have been tabulated, and are appended in as compact a form as possible. To accomplish so much work as is here represented in a single season, required the cordial coöperation of a considerable number of persons. It is proper that the names ·of those officers and students of the College who have faithfully and intelligently labored to accumulate these facts should be announced in connection with what they have done. If all who enjoy the privileges of students in natural science would exhibit the same enthusiasm for the acquisition of new truths, they would thereby not only improve themselves but increase the common stock of knowledge with a rapidity altogether unprecedented.

Prof. Levi Stockbridge has made nearly all the observations upon the flow of sap in the sugar maple, and has faithfully kept the record of the variations of pressure in the mercurial and water gauges on the sugar maple, the red maple and the butternut, which have been noted three or more times daily for several months.

Prof. S. T. Maynard has devoted much time to the care of the squash whose unparalleled performances in harness attest unmistakably its health and vigor. He has also kindly assisted in the preparation of gauges, and in every way in which his services were needed. The drawings for the cuts representing the squash and the apparatus used in the experiments with it, as well as for those relating to the specimens of elm, were furnished by him.

For the very convenient form of stopcocks used in the mercurial gauges, we are indebted to the ingenuity of Prof. S. H. Peabody.

Much credit is due to Mr. D. P. Penhallow, a post-graduate student, for his untiring devotion to the study of the squash-vine, with which he spent many days and nights, observing its mode of growth and making complete microscopical drawings of all its structure. He also adjusted gauges to several herbaceous plants, and reported upon the pressure of their saps. He assisted in finding the per cent. of water in various species of wood at different seasons of the year, and his pencil prepared all the drawings, except those already mentioned.

Charles Wellington, B.S., assistant in the chemical laboratory, has undertaken to determine the composition of various saps and the effect on them of the advancing season. This important investigation is not yet completed.

Mr. Walter H. Knapp, with great fidelity, furnished the material for the table showing the amount of sap which flowed daily from each species.

Mr. Atherton Clark made the observations on the water gauges, except that on the sugar maple, on the mercurial gauges in the case of the white birch root, the apple root and the three on the grape vine, one of which was thirty feet from the ground. He also did much of the work relating to the time when each species begins to flow.

Mr. William P. Brooks began and carried out very thoroughly a series of observations to learn precisely what species flowed, at what time in the season, and how rapidly, visiting for this purpose about forty species daily for several weeks. In some unaccountable manner, the memorandum book containing most of his records has been lost, and so his report is incomplete.

Mr. Henry Hague recorded the variations on the mercurial gauges upon the four birches, one of them thirty feet from the ground; and on the hornbeam, three times daily for many weeks.

Mr. George R. Dodge attended to a series of experiments

instituted to determine the circumstances which affect the flow of sap from the maples, and furnished an excellent report.

It has been said that all species of flowering plants will probably bleed from some part, if wounded, at some time of their growth. This has not been demonstrated, and some trees seem to have a wood so remarkably spongy and retentive of moisture as to render it unlikely that they should ever flow. Much effort has been made to arrive at the truth on this subject concerning our common forest trees by methods detailed below.

About the middle of last March, a large number of trees were selected and prepared for observation by boring one half-inch hole to the depth of two inches into the wood, and inserting a galvanized iron sap-spout, invented by Mr. C. C. Post, of Burlington, Vt., and well adapted for use in the sugar-bush. The species thus tapped, and all others named in this paper, will be mentioned by their common English names, which are familiar to most persons; but, in order that these may be clearly understood, a list is appended containing both the English and the Latin names. The following were tested, as above described, for sap, viz.: hemlock, black spruce, balsam fir, alder, European alder, striped maple, red maple, sugar maple, shad-bush, white birch, black birch, yellow birch, paper birch, hornbeam, chestnut, hickory, bitternut, cornel, thorn, quince, ash, beech, butternut, black walnut, mulberry, ironwood, white pine, yellow pine, buttonwood, aspen, English cherry, black cherry, mountain ash, apple, pear, peach, white oak, red oak, glaucous willow, white willow, bass, linden, elm and grape. These trees were visited every day, about noon, for several weeks, the holes being renewed as often as necessary, and whenever they were found flowing the number of drops per minute was recorded, except in the case of such trees as flowed somewhat abundantly and for a considerable time. The whole amount of sap from those of the latter class was carefully collected and weighed daily. A reference to the table appended will give at a glance the principal results, such as the dates of the beginning and end of the flow, and the total amount from each species. It will be seen that the sugar maple flows at any time when stripped of its foliage, provided the weather is favorable, the principal condition being a temperature above freezing, directly after severe frost. A comparison of the flow from this species with the pressure on the mercurial gauges, and with the temperature as indicated in the meteorological observations, kindly furnished by Prof. E. S. Snell, LL.D., of Amherst College, will convince the inquirer that there is an intimate connection between these three sets of facts.

The quantity of sap from a sugar maple during the season is much greater than from any other tree flowing from the same causes. Thus the entire flow from the butternut was less than the product of the sugar maple for a single day. The ironwood and the birches, however, surpass even the maple, both in the rapidity and amount of their flowing, if we make allowance for the difference in the size of the trees tested. A paper birch, fifteen inches in diameter, flowed in less than two months one thousand four hundred and eighty-six pounds of sap; the maximum flow, on the fifth of May, amounting to sixty-three pounds and four ounces, which is probably three times the average yield of a sugar maple of the same size. These latter species will not bleed during the winter, and seem to do so in the spring from a cause entirely different from that which affects the trees which bleed in fall and winter. The grape, which is often thought to bleed more freely than any other species, though later in the season, really flows but little, the total amount from a very large vine being eleven pounds and nine ounces.

Among the species subjected to trial, only those mentioned as bleeding exhibited this phenomenon. The following flowed for a short time, or very irregularly, or very slowly. The shad-bush was seen to flow, on the eighth of April, one drop in fifty seconds. The hickory bled one drop per minute of very sweet sap, on the fifteenth of April, and the cornel, ten drops on the same day. The European alder flowed three drops per minute, April ninth, and the common alder, four drops, on the twenty-first of March, and on the tenth of April, nine drops from one spout and six drops from another, inserted six inches below the former. The black walnut yielded a small amount of sap during several weeks, and, March thirtieth, bled six drops per minute. The buttonwood flowed forty drops per minute, March twenty-fifth, and one hundred on a very cold day, the eighth of April. The total amount, however, was very small. The apple bled twenty-eight drops per minute, May thirteenth, and the beech, on the tenth of May, flowed ten drops per minute, both yielding most sap in decidedly warm weather, the mean temperature for the last date being above 70° F. The latex of the mulberry exuded from the bark, on the ninth of April, as a transparent fluid which soon became milky, and the white and yellow pines flowed a small quantity of turpentine, apparently from both bark and wood.

A large red maple, which was thoroughly girdled in 1873, and whose bark had died and peeled off below the girdled place, was tapped above and also below it. The result was that it bled freely from both holes on many occasions. The flow, on the eighth of April, was fifty drops per minute from the upper one, and one drop from the lower one, while on the eleventh of the same month, it was three drops from the upper and fourteen drops from the lower one.

After the usual run of sap for the season has ceased, some species will bleed from the stump, if cut down, just like many herbaceous plants. Thus, Mr. Wm. F. Flint reports that large trees of the black, yellow, and paper birch, when felled on the thirtieth of June last, did not bleed immediately, as in April, but after an hour or two began to exude sap freely.

On August twenty-eighth, twenty-four species of young trees were cut down, about one foot from the ground, to see whether they would bleed. None did so immediately, but fifteen hours afterward the black birch ran a few drops, and the following were moist on the top of the stump, viz.: alder, yellow birch, red maple, cornel, ironwood, apple, elder, elm, and white pine. August thirty-first, the black birch bled a little, and the yellow birch, thorn, apple, glaucous willow, elm, and white pine were moist. The rest, including hemlock, shad-bush, white birch, chestnut, hornbeam, beech, ash, witch-hazel, bird cherry, white oak, red oak, and aspen, were perfectly dry, though all were sheltered from the sun.

These results seem to include most of the important attainable facts in regard to the flow of sap as exhibited by our common exogenous trees, and, while none of the observations can be exactly repeated from the nature of the phenomena, yet they may safely be accepted as the substantial truth concerning the whole subject.

The interesting facts observed last year, in connection with the attachment of mercurial gauges to the roots and trunks of trees which were known to bleed from wounds, and the suggestions derived from them, were a powerful stimulus to further investigations in this direction. Accordingly, a large number of gauges were prepared in early spring, and, as soon as the weather was suitable, attached to such trees and roots as gave promise of the most valuable results.

There still remained the unaccountable fact that the larger number of trees and shrubs did not show any tendency to bleed in spring, and therefore could not be made to answer any inquiries put to them in regard to the circulation of sap. It was thought best to adopt a cheaper and simpler form of gauge for application to such species as gave small promise of useful results. For this purpose, the following economical apparatus was devised and applied to the roots of elm, ash, white oak, chestnut, apple, sugar maple, and hickory. A straight glass tube, three feet in length, with a bore about one quarter of an inch in diameter, was joined by a conical rubber connector with each of the detached roots, and the roots again covered with the earth in which they grew. The tubes were now fastened in a vertical position to stakes set near the ends of the detached roots, which were one inch in diameter. They were then filled with water to a certain point, which was carefully marked, and the changes occurring noted every day. Sometimes the water in a tube would sink away, showing an absorption of the fluid by the roots; and again it would rise and flow over the top of the tube, demonstrating the fact that the absorbing power of the root was, sometimes at least, in excess of the affinity of the cellulose of the wood for water. It was well established that the wood of the roots of trees is in a condition in early spring to absorb with avidity the water from the tubes, while later in the season many of them exude water freely, so as to cause the tubes to overflow. The amount of absorption was recorded in inches, the minus sign being prefixed to the numbers, while the exudation was meas-

PHENOMENA OF PLANT-LIFE. 49

ured in a similar way, with the omission of the sign. Thirty-six inches of water in one of these tubes weighed one ounce, and from these data it was easy to learn the actual amount of water which was taken up or thrown off daily by each species. The table of observations for six common trees, which is appended, will convey a correct impression of the peculiarities of this phenomenon.

One of the most remarkable discoveries, in this connection, is the entirely unexpected fact that the roots of the sugar maple do not exude any sap from their wood when protected from frost, and show less independent power of absorbing water from the soil than almost any other species. Hence, there was no flow from the root into the tube at any time, but a constant moderate absorption of water from it.

The flow from the root into the tube is similar to that observed in the tube of an ordinary osmometer; but this does not prove that osmose has any influence in this matter, and the doubt about it is not diminished when we see the water moving, sometimes in one direction and sometimes in another. In the sugar maple, the flow was always out of the tube into the wood of the root; in the white oak, the absorption from the tube was, in some cases, as much as one ounce in thirty minutes, but rarely the current was reversed and absorption occurred from the ground; while, in the elm, the absorption from the tube was at its maximum, April fifteenth, and then gradually diminished until April twenty-first, from which date the flow into the tube continued till June thirtieth, when the observations were suspended.

A section of a white oak root, eight inches long and one inch in diameter, which was freshly dug from the damp earth, April eleventh, and weighed, was then placed with one end in water three-eighths of an inch in depth, and in ten hours absorbed 3.19 per cent. of its weight. This shows that the tissues were far from saturated, and were in an excellent condition to facilitate ordinary root absorption. A mercurial gauge attached to a root of white oak showed on the twelfth of April a suction sufficient to sustain a column of water 10.20 feet in height, which was caused by the absorption of the water in the connecting tube between the gauge and the root.

The mercurial gauge, which was used for determining the

variations of the pressure exerted by the sap of such species as are noted for the abundance of their flow, consisted of a syphon-tube of thick glass, the two legs of which were eight feet long, and about four inches apart. This was inverted and attached to a support of inch board, on the centre of which was fastened a scale divided to tenths of an inch. To one leg of the tube at the top was adjusted a brass stopcock, by means of small rubber hose, and to the stopcock was connected by a brass coupling a piece of thick lead pipe of small bore and convenient length, which was joined by another stopcock to the trunk, root, or branch which was to be tested. The stopcocks were so made, with a tube on the top, that communication could be opened between the free air and either the lead or the glass tubing at pleasure, and, when closed from the air, the passage was open between the mercury in the syphon-tube, the water in the lead pipe and the sap in the tree. The object of this three-way cock was to facilitate filling the tubes with water and mercury, and allowing the escape of any gas which might find its way into the apparatus from the tree. A sufficient quantity of mercury was poured into the inverted syphon to fill the two legs to the height of about forty inches, and the remainder of the leg connected with the tree, as well as the lead pipe, was carefully filled with water, all air being excluded. The other leg of the syphon-tube was left open to the atmosphere. When the sap exerted a pressure, it was indicated by a depression of the mercury in the closed leg of the glass tube and a rise in the open end, the difference between the two columns showing the pressure in inches of mercury. Suction into the tree was marked by the rise of the mercury in the closed leg and its depression in the open one, and in making the record the minus sign was prefixed to the figures expressing the number of inches of mercury.

One of the difficulties encountered in these experiments arose from the liability to leakage, either around the stopcock inserted into the tree, or from accidental wounds to the bark or small branches. In cases where the pressure was very great, it was sometimes necessary to solder a heavy sheet of lead to the stopcock and nail it to the tree with a packing of white lead in oil. Much trouble was also experienced from

PHENOMENA OF PLANT-LIFE.

the bursting of the lead pipes and the breaking of the glass tubes during severe cold weather by the formation of ice within the gauges. To avoid this as much as possible, the gauges were enclosed in wooden cases, and the more exposed portions wrapped in woollen blankets.

Mercurial gauges were attached to the following species, viz.: sugar maple, red maple, black, yellow, white and paper birches, ironwood, apple and grape, and all the observations may be found in the appended tables. The general results correspond with those of last year, but are much more complete, especially in regard to the two species which exhibit the most surprising phenomena and in which the public feel the deepest interest, namely, the sugar maple and the grape vine.

As soon as the discovery was made, by means of the water gauge, that the apple would flow from the root, a mercurial gauge was attached to a root an inch in diameter. At first, on the fifteenth of May, there was a slight suction amounting to -1.59 feet of water; but the pressure soon began, and rose to its maximum, May thirty-first, when it equalled 15.07 feet of water. Thus, the extreme variation was 16.66 feet.

The butternut had a range of only 13.03 feet, the minimum, -0.79 foot, occurring on April tenth, and the maximum, 12.24 feet, on April fourteenth.

The red maple attained its minimum, -2.83 feet, April sixteenth, and its maximum, 18.59 feet, April eighth, the total variation being 21.42 feet of water.

The ironwood exerted its greatest suction on the nineteenth of May, which equalled -24.60 feet, while the greatest pressure was 40.35 feet, and was observed, May thirteenth. The total variation was thus 64.95 feet of water.

The white birch began early in the season, April ninth; reached its minimum, -19.26 feet, on the eleventh of May, and its maximum, 39.66 feet, April twenty-third. The extreme variation was, therefore, 58.92 feet of water.

A gauge was attached to a root of white birch on the eighth of April; the pressure began, April twelfth, and steadily advanced to its highest point, 38.08 feet, May twelfth, and declined to zero, May twenty-third, and to its minimum, -22.98 feet, August twenty-sixth, the extreme variation amounting to 61.06 feet of water. The root was dug up in October and found apparently alive and healthy.

The black birch root last year exerted the astonishing pressure of 84.77 feet of water, but was not observed through the season. This year, on the eighth of April, a guage was adjusted to a root of the same tree, and, although the pressure was not quite as great as last season, the extreme variation was 102.68 feet. The first pressure was, April twenty-third, and the highest, May tenth, and equalled 77.06 feet, while the greatest suction was on September fourteenth, and amounted to −25.62 feet of water.

The pressure is evidently caused in these roots, which are entirely detached from the tree and lie in the earth just as they grew, by the activity of their power of absorption, which seems to be greatest just as the buds are about bursting. The suction is remarkably powerful, and must apparently result from some chemical change occurring in the root, after the root-fibres have lost their absorbing power. A critical examination by the chemist and the microscopist would probably give an explanation for this phenomenon.

The paper birch tree reached its maximum, May sixth, when the pressure was equal to sustaining a column of water 61.20 feet in height. The suction on June fourteenth was −7.93 feet, and the extreme variation for the season was 69.13 feet.

On the eighth of April, a gauge was attached to a yellow birch tree near the ground, and, on the twenty-fourth, at noon, the pressure was 73.67 feet of water. A hole was then bored into the tree at a height of thirty feet above the lower one, for the purpose of putting up another gauge. The mercury in the lower gauge fell at the rate of four inches per minute, till it stood at a point representing 35.13 feet of water. The sap, at the same time, flowed freely from the upper orifice. The usual difference between the gauges thus placed thirty feet apart was from twenty-four to thirty-five feet of water, showing evidently that the power furnishing the pressure was from below, that is, from the root. The maximum of the lower gauge was 74.22 feet, April twenty-second, and the minimum was −22.44 feet, May sixteenth, and, hence, the total variation was 96.66 feet. The upper gauge attained a pressure of 41.25 feet, on the ninth of May, and sank to −11.11 feet on

PHENOMENA OF PLANT-LIFE. 53

the thirteenth of May, the extreme variation being 52.36 feet of water. After the development of the buds, the upper gauge stood uniformly at from −1 to −4 feet of water, and the lower one was mostly minus.

The bleeding of a broken grape vine, in 1720, induced the Rev. Stephen Hales, an ingenious observer of nature, to attach mercurial manometers to the stumps of vine branches and stems, by means of which he obtained a maximum pressure of forty-three feet of water. These experiments were made on vines of the species *Vitis vinifera*, in the comparatively cool and moist climate of England. It is, therefore, not surprising, that the more vigorous *Vitis æstivalis*, in the more fervid and sunny climate of Massachusetts, should exert a greatly superior force. In order to determine as many facts as possible concerning the flow and pressure of the sap of the wild summer grape, two of the largest vines on the College estate were selected and prepared for observation. The smaller one was about three inches in diameter at the ground, and spread over a young elm, some forty feet in height, and standing in moist, open land. One of the main roots of the vine was uncovered and followed from the stem toward its extremities, a distance of four feet, where it was cut off. To the large end of this detached root, the remainder of which was left undisturbed in the soil as it grew, was firmly fastened a piece of stout rubber hose, which was connected by means of a stopcock to the lead pipe of a mercurial gauge. This was on May-day. The tissues of the root, which had not yet awakened from its winter sleep, at once began to absorb the water from the gauge, and the next day there appeared a suction equal to −4.53 feet of water. This continued, though gradually diminishing, till it reached zero, on the tenth of May. From this time the pressure still increased until, on the twenty-ninth of the month, it became sufficient to sustain a column of water 88.74 feet in height, which is more than twice as great as the maximum observed by Hales, and the greatest pressure ever produced by the sap of a plant so far as we know. It is an interesting fact that this maximum occurred on the warmest day in May, the mean temperature having been 71.7° F. It is also noteworthy that, on the very day when the gauge first showed pressure, the vine which was tapped began to flow, though it was half a mile distant. The pressure on the gauge steadily diminished through the season, and, on the fourteenth of September, amounted to 19.35 feet. The extreme variation was 93.27 feet of water, and, therefore, 9.41 feet less than in the case of the black birch root, which exhibited a much greater suction, though less pressure, than the grape root.

The other vine selected for trial was nearly four inches in diameter and more than fifty feet high. To a large branch of this, near the ground, was attached a gauge by means of a rubber hose, the branch being cut off for that purpose. A second gauge was secured to another branch at the height of thirty feet above the first, and observations made upon them, once, twice, or three times, daily, from May seventh till June thirtieth. After this, occasional visits were made to the vine, though the variations were very slight. The pressure on the lower gauge began on the seventh of May, when it was 11.11 feet of water, and reached its maximum on the twenty-sixth of the month, equalling a column of water 83.87 feet in height. The pressure declined quite rapidly as soon as the buds began to develop, and fell to zero, June thirteenth. The greatest suction was exhibited on the twenty-ninth of June, and was equal to sustaining a column of water 14.39 feet high. During the month of July, when growth was most rapid, the suction was uniformly about -7.37 feet of water, and, during August, about -4 feet. The extreme variation on this gauge amounted to 98.26 feet, though the pressure was somewhat less than was shown by that on the detached root of the vine already mentioned.

The upper gauge was not reached by the sap rising from the root until some days after pressure was manifest at the lower one. On the twelfth of May, the lower one stood at 34.11 feet of water, and the upper at 3.40 feet. The maximum pressure was attained, May sixteenth, and was 39.66 feet, while the greatest suction occurred, June twentieth, and was -10.77 feet. The extreme variation of the upper gauge was 50.43 feet. The difference between the two gauges was usually from 20 to 30 feet of water; but when the pressure on the lower one was greatest, the difference was 60.41 feet, in consequence of the fact that the force was entirely from the root,

PHENOMENA OF PLANT-LIFE.

and the wood of the vine was a hindrance to the sudden upward thrust of the sap. After the foliage was developed, the suction was limited to from −6 to −12 feet of water, on account, doubtless, of the porous character of the foliage and young branches, and there was no great difference between the gauges.

The flow of sap from the sugar maple, so familiar to all, and yet so variable and peculiar, was the first object of investigation in the beginning of these experiments, in 1873, but its mysterious fluctuations were not fully known nor understood until the close of the year 1874. The extraordinary facts, that the flow occurred in mid-winter and early spring, when the ground was covered with snow and there were no signs of life; that the flow began only during mild days immediately following a severe frost, and ceased usually after a few hours; that when a cavity was cut into a sugar maple tree, the sap flowed down from above, while in a birch it flowed most freely from below; and especially the fact, that when a gauge was attached to a tree, it exhibited the most surprising variations from great pressure, during the day, to powerful suction at night,—these, and other unaccountable things, seemed to demand special effort to discover all the phenomena attending the flow of maple sap; and then, if possible, to invent some rational explanation of them.

Accordingly, a large number of experiments were devised and carried out, with a very great amount of labor and no little expense. Among them were the collection and weighing of all the sap which would flow from a healthy tree, from November to the following May, with a careful observation of the times when the flow began and ceased, in each case of good sap-weather; the collection, weighing and analysis of sap during different periods of the entire season, both from the usual level and from the top of a tree thirty feet from the ground; the collection and examination of the gas which escapes with the first flow of sap from the orifice first made in a tree in the spring; the effect of increasing the number of holes upon the total flow of sap and the entire product of sugar; the result of tapping trees at various elevations from the earth, on different sides, and to different depths; and finally, a record for comparison and study of the fluctuations in the mercury of several gauges, attached to various parts of the same tree, as observed three or more times daily.

Upon reference to the table showing the flow of sap from the sugar maple, it will be noticed that the tree (No. 1) tapped near the ground flowed quite freely in December and January as well as in March and April, the total amount of sap being five hundred and sixty-six pounds and twelve ounces. Notwithstanding the large quantity previously exuded, the flow from this tree during the month of April amounted to one hundred and four pounds and eight ounces, while a tree (No. 2) nearly as large, from which no sap had been taken, but which was tapped at the height of thirty feet from the ground, bled only fifty-five pounds and eleven ounces. It is evident, therefore, that the flow is greatest at the lowest point, other things being equal; but it often happens that the sap will drop from a broken twig in the top of a tree when it will not run at all from the trunk.

Mr. Samuel F. Perley, of Naples, Maine, in an interesting communication containing much valuable information derived from his large experience in the sugar-bush, relates the following incident: "Happening, on a bright, sunny morning, to visit a sugar tree standing in open land, and having a large, spreading top, I was surprised, on walking beneath the limbs, to find quite a smart shower falling upon me. On looking up, I could see no clouds, yet the drops were falling thick and fast in all the area covered by the branches of the tree. An examination showed the drops to be drops of sap flowing from innumerable broken twigs. I then remembered that a day or two before there had been a storm of sleet and rain, which had encased the trees with a heavy coating of ice, and following that, a violent wind which had twisted and broken many of the smaller branches. From these was now flowing a brilliant shower of sap, sparkling in the bright sunshine. I could not perceive that this wholesale tapping diminished at all the flow from the trunk, or in any manner injured the tree."

Icicles of frozen sap are not unfrequently seen depending from the branches of maple and butternut trees during severe cold weather, when the temperature rises only slightly above 32° F. at mid-day. On Thanksgiving Day, 1874, the

PHENOMENA OF PLANT-LIFE.

thermometer, in the shade, indicated 32° F. at two P. M.
A sugar maple was tapped at the ground, and fifty feet above it, and while there was no flow from the lower orifice, the upper one bled four drops per minute.

On the twentieth of November last, the weather was cold, and at eleven, A. M. there was a rapid fall of soft snow, followed by a rising temperature. At half-past twelve, P. M., the mercurial gauge, in the top of a sugar maple, indicated a pressure of about nine feet of water, while a gauge at the ground showed neither pressure nor suction.

In the case of a tree tapped in 1873, on the north and south sides, in order to compare the flow from each, it was found that, for some reason, the north spout yielded nearly twice as much sap as the south one, and flowed two weeks longer. It appears probable that this was an exceptional instance, and possibly to be accounted for by the fact, that the roots of the south side ran under a highway, while those of the north side luxuriated in a rich meadow.

In 1874, another tree, about sixty feet in height and four feet and ten inches in girth, was subjected to the same trial. The total flow from the south side was eighty-six pounds and four ounces, while that from the north side was sixty-eight pounds and five ounces. Near the close of the season only, did the flow from the latter exceed that from the former. There can be no doubt that it is much wiser to tap all sugar trees on the south side, because the sap will flow earlier and more abundantly than from the shaded side, while the late sap is of little value to the sugar-maker.

Another sugar maple, seventy feet high and four feet in circumference, was tapped on the south side in five places, the holes being two feet apart on a vertical line, so that spout number one was near the ground, number two, directly above number one, number three, two feet above number two, and so on. During the month of April, the sap from each spout was weighed daily, and the results were as follows, viz.: The total flow was one hundred and twenty pounds and one ounce. From number one, near the ground, was collected seventy-eight pounds and ten ounces; from number two, twelve pounds and two ounces; from number three, five pounds and ten ounces; from number four, eight pounds and seven ounces; and from number five, fifteen pounds and four ounces. These facts are, in the main, what would be expected from the other observations made concerning the flow of maple sap.

The effect of increasing the number of spouts inserted into a tree was tried on two red maples, which flow much less than the sugar maple and for a shorter time. Ten spouts in one tree, sixty feet high and four feet eight inches in girth, were found to flow, during the first half of April, seventy-eight pounds and eight ounces, while one spout in a similar tree flowed less than half as much, or thirty-five pounds and two ounces. There can be no doubt that the quantity of sap obtained from a tree by the use of many spouts is greater than that from a limited number, but it is not likely to contain so large a per cent. of sugar. Still, if it be true, as seems probable, that the withdrawal of sap exerts no deleterious influence upon the health and vigor of a tree, and the sap is richest early in the season, it would seem best to insert more spouts, and so extract the sugar in its purest condition as rapidly as possible. This, of course, would necessitate a greater expenditure for buckets, which might possibly counterbalance the advantages of the new method. Experiments might be easily instituted to determine the facts in regard to this matter by any intelligent sugar-maker.

In regard to the origin of cane sugar in the sap of the maples, the butternut and the black walnut, we must, for the present, admit that we have not yet discovered it; though the singular fact that the species which yield this sugar belong to that class of trees which only flow freely after severe frost seems to indicate that freezing and thawing may have some influence upon its production.

It will be seen, from an examination of the table relating to the composition of saps, that the sap of the wild grape is almost pure water, and that it contained, on the fifteenth of May last, no trace of either cane sugar, glucose or starch. There is, however, in the wood of the roots and stems of the genus *Vitis* a great quantity of a colorless, translucent, almost tasteless mucilage, which is abundantly exuded from the pores of a cross section made at any time when the roots are dormant. Very little even of this seems to escape from a bleeding vine, which may account for the fact that the flow of crude sap from the grape does not perceptibly affect its subsequent growth or productiveness.

The sap of the sugar maple contains from two to three per cent. of cane sugar, while that of the red maple yields only about half as much. The sap of the latter is said by Mr. H. M. Sessions, of Wilbraham, also to contain some ingredient which attacks iron, forming a very dark-colored syrup when evaporated in pans of that metal. It is, therefore, better to exclude it from the sap gathered for the manufacture of sugar.

In order to obtain as much information as possible in regard to the sap of the sugar maple, an analysis was made of the gas contained in the tree when first tapped. This was procured by inserting a stopcock into the sap-wood of a tree twenty feet from the ground. To the stopcock was attached a glass tube by means of a rubber connector and the tube passed through a cork into a large bottle, reaching to the bottom. As soon as the bottle was filled with sap, it was tightly closed and taken to the laboratory, where the gas was separated by boiling. The analysis shows that the gas contains much less nitrogen and more oxygen than atmospheric air, while the proportion of carbonic acid gas is about one hundred and thirty-four times greater in the former than in the latter.

As we do not know how or when the cane sugar is formed in the maple, we cannot account for the variations in the sweetness of its sap, which are, however, very great. As the flow depends upon the freezing and thawing of the wood, and possibly upon the continuance of absorption by the roots to supply the drain upon the tapped tree, it is evident that a large body of snow upon the ground will favor it, since the earth will then be warmer and the night temperature of the air much colder than under other circumstances. It does not appear that there is any greater proportion of sap in the maple than in many other trees, but only that for some unknown reason it is separated in greater quantity by freezing, or else not reabsorbed after such separation so quickly as in other species.

For the purpose of learning whether root absorption is necessary to keep up the flow of sap through the season, a large tree, sixty feet in height and four feet and a half in girth, was cut early in December, 1874, and firmly lashed in an upright position to neighboring trees. A fire was then kindled around the lower end of the trunk, in order to dry and close as far as possible the pores of the wood. Next spring it is proposed to apply mercurial gauges to determine whether the sap moves, as in trees in a natural condition, and afterward to collect and analyze the sap.

While it is certain that the flow of the grape and the birch results from the great activity of the absorbing rootlets when they first awake in spring from their winter's repose, it seems equally evident that root absorption has no direct connection with the flow of maple sap. This discovery was made by means of five mercurial gauges, which were attached with great care to a fine, vigorous tree, about sixty feet in height, on the twentieth of last March. The gauges were so connected with all parts of the tree that every movement of the sap would be indicated. Number one was joined to a stopcock inserted into the sap-wood about two feet from the ground, the hole being about one inch in diameter and two inches deep. Number two was connected by a stout rubber hose to a root one inch in diameter, which was laid bare by the use of a force-pump, so as to avoid breaking any of its fibres. This root was cut open at the distance of about two feet from the tree, and gauge number two united to the stump, which was attached to the trunk. Number three was joined in the same way to the large end of the detached root, which remained in the soil just as it grew. Number four was fastened to a piece of gas-pipe one inch in diameter, which was screwed into the tree to the depth of ten inches, a thread having been cut for this purpose on the outside of it. No sap could enter this gauge except at the very centre of the heart-wood of the trunk. Number five was attached to the sap-wood among the branches, at an elevation of twenty feet above gauge number one. The gauges thus connected were then inclosed in tight pine cases, and the metallic pipes and stopcocks wrapped in woolen blankets to protect them from the cold. The observations were taken regularly at six A. M., at noon, and at six P. M., for about ten weeks, until the changes became unimportant. The table appended gives all the variations of sap pressure in different parts of the tree, as recorded at the times specified. A reference to figure 44 will convey a correct idea of the manner in which the mercury fluctuates during every hour of the day and night.

The following are some of the most interesting results obtained from the several gauges:—

GAUGE.	Minimum.	Date of Minimum.	Maximum.	Date of Maximum.	Extreme Variation.
Gauge 1,	—18.13	Apr. 11,	39.67	Mar. 28,	57.80 feet of water.
" 2,	—7.71	" 4,	36.27	" 28,	43.98 " "
" 3,	—7.71	Mar. 21,	3.40	Apr. 3,	11.11 " "
" 4,	—6.01	Apr. 22,	22.33	Mar. 28,	28.34 " "
" 5,	—26.07	Mar. 31,	52.13	Apr. 2,	78.20 " "

The wood of the detached root absorbed the water from the gauge, so as to exert a suction, like the roots of most other species of trees in early spring, but the pressure exhibited at any time was scarcely worthy of mention. So strange did this appear, that, on the fourth of April, the gauge was removed to a healthy root, detached from another tree, and, to avoid any possibility of error, it was afterward connected with a third root, but the results were always similar. It is certain, therefore, from these observations, as well as those connected with the water-gauge, described on a preceding page, that the rise and flow of maple sap is not directly caused by the activity of absorbent rootlets.

Secondly; it is seen that the movements of the sap in the heart of a tree are much less rapid and vigorous than those occurring in the sap-wood at the same level. This is doubtless owing to the fact that the old wood is more dense, and therefore less permeable to fluids than the outer layers of alburnum; and also to the circumstance that the variations of temperature, at the depth of ten inches from the bark, are necessarily slow and limited.

Finally; it remains to consider the extraordinary fact, that the greatest suction, as well as the highest pressure, was exhibited by the gauge in the top of the tree. On the eighteenth of April, the lower gauge in the sap-wood indicated a pressure equal to 10.77 feet of water, while, at the same time, the upper gauge showed a pressure of 24.93 feet. On the thirty-first of March, the gauges were all frozen, number one standing at 28.90 feet of water, while number five indicated a suction equal to −26.07, a difference of 54.97 feet. In the case of number one, attached to the trunk near the ground, it seemed that the gauge froze before the body of the tree was much chilled, while, by the sudden freezing of the branches, the sap was abstracted from the upper gauge before the cold had penetrated the coverings sufficiently to freeze it.

On the nineteenth of April, the upper gauge showed little or no pressure, while the lower one still indicated a pressure of 17 feet. This was apparently due to the absorption of the sap from the branches by the expanding buds.

In view of all the phenomena thus far observed, it appears that the flow of sap from the maple and other species, which bleed only after being frozen, is in no sense a vital process, but purely physical. The sap is separated from the cellulose of the wood by the cold, and, under ordinary conditions, gradually reabsorbed. If, however, the tree be tapped, so that the liberated sap can escape, then it will do so, flowing, as is readily seen to be the case with the maple, most copiously from above. The bleeding is, therefore, a sort of leakage from the vessels of the wood, but this is doubtless increased by the elastic force of the gases in the tree, which are compressed by the liberated sap, and this expansive power must be intensified by the increase of temperature which always accompanies a flow.

This theory explains the fluctuations of the gauges, and accounts for the singular fact that the upper one shows the most pressure and the greatest variations, inasmuch as the branches and twigs would, of course, be most quickly and powerfully affected by the heat of the sun and the temperature of the atmosphere. The pressure of the expanded gases in a tree in a normal condition would facilitate the re-absorption by the wood of the liberated sap. Their contraction by cold would also cause the cessation of the flow from a tree which was running, and produce the remarkable phenomenon of suction exhibited by the gauges at night or during frosty weather.

An important and elegant demonstration of this theory was obtained by cutting large branches, fifteen to twenty feet in length, when the thermometer was below zero, from trees of the sugar maple, white birch, elm, hickory, buttonwood, chestnut and willow, and suspending them in the warm air of the Durfee Plant-House. The maple soon began to bleed at the rate of twenty-four drops per minute, while the buttonwood bled eleven drops, and the hickory exuded a little very sweet sap, precisely as in spring. The birch, elm, chestnut and willow did not flow at all, and were not even moist on the freshly-cut surface.

A mercurial gauge, attached to the end of a frozen branch of sugar maple, indicated pressure and suction when the temperature was raised and lowered, precisely as it would have done upon a maple tree during the ordinary alternations of day and night in the spring of the year when the sap is flowing.

In the warm regions of Asia, Africa and America, are found about one thousand species of palm trees, from many of which a sweet sap is obtained in large quantities. This is simply allowed to ferment, and drank as palm-wine or toddy, or distilled for the production of a sort of brandy, or it is evaporated for the extraction of its sugar in the form of syrup, or of a more or less crystalline solid called jaggery. In the province of Bengal, in India, more than one hundred million pounds of palm-sugar are manufactured annually, while the total product of palm-wine in the world greatly exceeds that of wine from the grape.

There are three principal methods adopted in different countries for obtaining the sweet sap of palms. In Chili, trees fifty feet high are felled in such a way that the top will lie higher than the butt of the trunk, and the single terminal bud with the crown of leaves is cut off. The sap flows abundantly from the higher end of this log, and if a fresh slice of wood be removed every day the bleeding continues for several months. The yield is greatest during the warmest days, and amounts in all to an average of ninety gallons, or about seven hundred and twenty-five pounds, from each tree. This sap is mostly evaporated and utilized as a very agreeable syrup called palm-honey.

In India, it is customary to make incision into the wood of trees near the top, from which, during the cool months, the sap flows freely. From the common wild date-palm the annual yield of sap is about two hundred pounds, containing some eight pounds of sugar, or four times the average product of the sugar maple. Much the larger proportion of palm sap is obtained, however, from the large branching flower-stalks of the inflorescence. These are produced in the axils of the immense leaves or fronds, and before they burst the spathe in which they are enveloped, they are carefully bound together with pieces of palm-leaf. These buds are then beaten every morning with sticks and a thin slice removed from the tip of the axis of inflorescence. From the freshly exposed surface the sweet sap runs very abundantly for several months. Indeed, some species continually send out new flower-stalks, which are constantly bled until, after two or three years, the tree dies from exhaustion.

But the most remarkable flow of sap is that of the *Agave Americana*, or century plant. This is the largest herbaceous plant known, the leaves of one in the Durfee Plant-House being eight feet long and of immense weight. In Mexico, the sap of this species furnishes the favorite beverage of the people. This is called pulque, and has a most detestable odor of carrion and a slightly acid taste. The Mexicans are very fond of it, and natives of other countries soon learn to love it and then prefer it to claret. The sap is procured by cutting out the bud of the inflorescence which appears in the centre of the massive crown of leaves, and, if undisturbed, develops into a flower-stalk from thirty to forty feet high and covered with thousands of blossoms. The cavity made by removing the bud is speedily filled with a sweet sap, and the total amount from one plant is stated by Von Humboldt to be from twelve to sixteen hundred pounds. The plant then dies from exhaustion.

It is impossible to give any satisfactory explanation for these extraordinary phenomena. It is easy to state that these plants produce large quantities of starch and sugar preparatory to flowering, but why should they continue to flow so long after the trees are cut down or the flower buds removed?

PHENOMENA OF PLANT-LIFE.

If it be true that the sap of plants flows to the points of consumption, it is still difficult to explain why it should persistently tend upward to the top of a prostrate trunk, or of a standing tree, for months after the bud, for the special nourishment of which it is designed, has been destroyed, and after the process of growth has been entirely suspended.

It is evident, in conclusion, that there yet remains ample room for investigation concerning the phenomena connected with the development of plants and the circulation of sap. Though we cannot hope to exhaust the subject, or to discover precisely what the force is which we call life, and which imparts to every species and individual of the vegetable world its peculiar form and characteristics, it is none the less important and interesting to exercise our utmost ingenuity in the effort to discover the times and modes of its operation, and its relations to the other forces of Nature.

LATIN AND COMMON NAMES OF SPECIES.

Latin	Common	Latin	Common
Abies balsamea,	Balsam Fir.	*Pinus longifolia,*	Chir.
A. Canadensis,	Hemlock.	*P. rigida,*	Yellow Pine.
A. nigra,	Black Spruce.	*P. Strobus,*	White Pine.
A. Picea,	Silver Fir.	*Platanus occidentalis,*	Buttonwood.
Acer Pennsylvanicum,	Striped Maple.	*Populus tremuloides,*	Aspen.
A. rubrum,	Red Maple.	*Prunus Amygdalus,*	Almond.
A. saccharinum,	Sugar Maple.	*P. Armeniaca,*	Apricot.
Alnus incana,	European Alder.	*P. Avium,*	English Cherry.
A. serrulata,	Alder.	*P. domestica,*	Plum.
Amelanchier Canadensis,	Shad Bush.	*P. Pennsylvanica,*	Bird Cherry.
Betula alba var. populifolia,	White Birch.	*P. Persica,*	Peach.
		P. serotina,	Wild Cherry.
B. lenta,	Black Birch.	*P. Virginiana,*	Choke Cherry.
B. lutea,	Yellow Birch.	*Pyrus aucuparia,*	Mountain Ash.
B. papyracea,	Paper Birch.	*P. baccata,*	Siberian Crab.
Carpinus Americana,	Hornbeam.	*P. Malus,*	Apple.
Castanea vesca. var. Americana,	Chestnut.	*Quercus alba,*	White Oak.
		Q. coccinea var. tinctoria,	Black Oak.
Carya alba,	Hickory.		
C. amara,	Bitternut.	*Q. rubra,*	Red Oak.
Cornus alternifolia,	Cornel.	*Salix alba,*	White Willow.
Cratægus coccinea,	Thorn.	*S. Babylonica,*	Weeping Willow.
Cydonia vulgaris,	Quince.	*S. discolor,*	Glaucous Willow.
Fagus ferruginea,	Beech.	*Syringa vulgaris,*	Lilac.
Fraxinus Americana,	Ash.	*Tectona grandis,*	Teak.
Juglans cinerea,	Butternut.	*Tilia Americana,*	Bass.
J. nigra,	Black Walnut.	*T. Europea,*	Linden.
Morus alba,	Mulberry.	*Ulmus Americana,*	Elm.
Ostrya Virginica,	Ironwood.	*Vitis æstivalis,*	Grape.
Phoradendron flavescens,	Mistletoe.	*V. vinifera,*	European Grape.

TABLE

Showing the date and amount of the Flow of Sap from species which bleed somewhat freely, with dimensions of specimens under observation.

NAME.	Height, feet.	Girth, feet. in.
Acer Pennsylvanicum,	22	1
A. saccharinum, No. 1,	60	6 10
A. saccharinum, No. 2,	60	5 6
Betula alba var. populifolia,	40	2 4
B. lenta,	57	4 2
B. lutea,	58	3 1
B. papyracea,	75	3 9
Carpinus Americana,	16	8
Juglans cinerea,	51	4 2
Ostrya Virginica,	54	2 2
Vitis æstivalis,	35	8

Total amount of Sap collected from the following species during the season of 1874.

DATE.	Pounds.	Ounces.
Acer saccharinum, from Dec. 16, 1873, to Apr. 30, 1874,	566	12
A. saccharinum, (30 feet from ground) from April 1, to May 1,	55	11
A. Pennsylvanicum, from Mar. 23, to May 4,	15	15
Betula alba var. populifolia, from Mar. 23, to May 23,	127	6
B. lenta, from Mar. 29, to May 29,	397	7
B. lutea, from Apr. 3, to May 27,	949	9
B. papyracea, from Mar. 29, to May 26,	1,486	–
Carpinus Americana, from Apr. 9, to May 22,	6	13
Juglans cinerea, from Mar. 23, to May 18,	18	13
Ostrya Virginica, from Apr 16, to May 26,	279	–
Vitis æstivalis, from May 11, to June 3,	11	9

PHENOMENA OF PLANT-LIFE.

Flow of Sap.

DATE.	Acer Pennsylvanicum.	Betula alba var. populifolia.	Juglans cinerea.	Betulapapyracea.	Betula lenta.	Acer saccharinum.	Betula lutea.	Carpinus Americana.	Ostrya Virginica.
1874.	lbs. oz.	lbs. oz.	lbs. oz.	lbs. oz.	lbs. oz.	lbs. oz.	lbs. oz.	lbs. oz.	lbs. oz.
Mar. 23,	5	1 2	10	–	–	–	–	–	–
26,	14	7	1 10	–	–	–	–	–	–
27,	4	6	1 5	–	–	–	–	–	–
28,	13	2	9	–	–	–	–	–	–
29,	1	2 11	7	3	1	–	–	–	–
30,	4	1 8	8	–	–	–	–	–	–
31,	13	2 4	1 1	2	–	–	–	–	–
Apr. 1,	2	2 2	5	2	–	2	–	–	–
2,	9	1	2	–	–	4 7	–	–	–
3,	5	2	10	1	–	7 15	2	–	–
4,	10	2 15	10	11	2	–	1 10	–	–
5,	–	1 2	–	10	1 1	1	2	–	–
6,	8	1	6	2	2	4	2	–	–
7,	15	4	11	4	10	5 2	1 12	–	–
8,	1 9	3 10	6	1 11	2 13	8 8	5 4	–	–
9,	3	7 15	1 3	4 14	6 7	7	8 5	1 3	–
10,	3	9 8	· 13	9 2	8	–	9 2	1	–
11,	7	10 12	5	9 9	11 4	1 4	13	3	–
12,	–	9 1	5	12 9	14 4	–	14 8	–	–
13,	6	–	1	1	12 4	1 4	6 2	–	–
14,	6	4	10	2 4	7 12	7 12	8 6	–	–
15,	1	6 10	12	14 13	16 11	1 10	18 14	–	–
16,	3	9 2	4	25 5	20 10	–	27	–	1
17,	1	10	–	28 9	22 2	–	25 7	1	2 15
18,	1	4 13	1	27 14	20 14	3 7	19 6	–	1 3
19,	1 10	7 1	15	38 1	20 4	8	31 4	–	–
20,	11	8 14	–	47 1	22 2	1 2	47	–	2 14
21,	5	7 10	11	57 2	12 10	8	54 9	1	2 1
22,	4	7 7	1	51 2	5 12	–	31 8	–	12
23,	4	7 3	5	53 13	5 14	–	21 5	–	14
24,	1	5 11	3	51 5	6 3	–	15 13	–	2
25,	3	5 15	2	58 7	6 4	–	13 6	–	12
26,	1	2 2	1	53 11	6 14	3 6	10 2	–	4
27,	15	3 5	9	56 2	5 7	–	8 15	–	1
28,	3	2 6	3	54 9	6 9	2 6	9 12	–	4
29,	5	2	5	52 12	5 3	–	9	–	–
30,	1 6	1 9	2	50 7	4 13	–	7 14	–	–
May 1,	1	1 15	1	52 2	4 15	–	8 8	–	–
2,	–	2 4	–	52 5	5 7	–	9 8	–	–
3,	–	1 15	3	57 5	14 7	–	24 11	–	–
4,	1	2 1	11	62 10	6 15	–	42 13	–	1 4

Flow of Sap—Continued.

DATE.	Vitis aestivalis.	Betula alba var. populifolia.	Juglans cinerea.	Betula papyracea.	Betula lenta.	Betula lutea.	Carpinus Americana.	Ostrya Virginica.
1874.	lbs. oz.	lbs. oz.	lbs. oz.	lbs. oz.	lbs. oz.	lbs. oz.	lbs. oz.	lbs. oz.
May 5,	—	2 4	7	63 4	5 15	47 1	—	6 2
6,	—	1 4	—	57 13	5 5	43 14	15	9 10
7,	—	1 7	3	53 14	5 8	42 7	4	12 4
8,	—	1 3	—	49 1	4 2	39 8	1	11 3
9,	—	1 4	—	46 7	3 12	37 8	—	9
10,	—	2 3	—	50 6	6 9	43 2	—	15 5
11,	6	1 8	—	36 1	8 8	33 8	4	22 3
12,	6	5	—	30 2	12 5	30 4	5	24 4
13,	3	8	—	26 6	11 14	26 6	1	24 2
14,	6	2	—	23 5	4 10	22 6	—	26 3
15,	12	—	—	12 10	2 15	10 6	—	20 10
16,	2 5	3	—	9 3	1 15	7 15	1 4	19 14
17,	2 6	1 4	—	14 10	13 13	15 10	1 9	27 2
18,	2 1	4	1	7 9	4 9	2 5	1	14 12
19,	1 8	—	—	3 14	11 9	8 11	7	14 13
20,	13	1 13	—	2 2	1 15	2 5	—	3 1
21,	8	—	—	7 1	12	12	—	1 14
22,	8	12	—	14	6 2	5 6	1	7 13
23,	4	—	—	—	—	12	—	5
24,	3	—	—	—	—	—	—	—
25,	6	—	—	—	—	—	—	—
26,	10	—	—	3	11	2 7	—	1 3
27,	8	—	—	—	—	—	—	—
28,	1	—	—	—	—	—	—	—
29,	1	—	—	—	—	—	—	—
30,	2	—	—	—	—	—	—	—
31,	1	—	—	—	—	—	—	—
June 1,	—	—	—	—	—	—	—	—
2,	—	—	—	—	—	—	—	—
3,	3	—	—	—	—	—	—	—

Flow of Sap—Concluded.

Date.	Acer saccharinum.	Date.	Acer saccharinum.	Date.	Acer saccharinum.
1873.	lbs. oz.	**1874.**	lbs. oz.	**1874.**	lbs. oz.
Dec. 16,	16 7	Mar. 2,	21 15	Mar. 30,	24
17,	—	3,	25 4	31,	6
18,	—	4,	8 13	Apr. 1,	5
19,	—	5,	11 8	2,	6
20,	8	7,	7 3	18,	36
21,	3	8,	23 3	19,	19
22,	10	14,	5 11	20,	—
		15,	17 4	21,	12
1874.		16,	29	22,	8
Jan. 2,	5 3	17,	23	23,	4
3,	8 10	18,	24 12	24,	3
4,	10 5	19,	16	25,	—
7,	—	20,	6 4	26,	11
8,	10 9	21,	48 4	27,	10 8
9,	10 1	22,	6 10	28,	—
10,	2 10	25,	12 8	29,	—
11,	4	27,	24	30,	—
23,	—	28,	25		
24,	2 15	29,	12		

TABLE

Showing the variations in Water Gauges attached to roots of trees. The figures indicate inches of water in tubes of such size that a column of thirty-six inches weighs one ounce. The minus sign denotes absorption of the water by the root, and the absence of the sign denotes flow of sap from the root. The size of the trees is unimportant, but they were all vigorous specimens, standing in open ground. A summary of the principal facts relating to the four species which showed the greatest fluctuations is given below.

Acer saccharinum.—Water gauge attached, May 1. Maximum absorption was 69 inches or 1.91 ounces of water, May 10. Minimum, 2 inches or 0.055 of an ounce, June 1. Total absorption, in the month of May, 410.7 inches or 11.4 ounces. No flow of sap.

Quercus alba.—Gauge attached, April 11. Maximum absorption, 46 inches or 1.28 ounces, May 2. The tube, however, was often emptied of its contents within an hour or two after it was filled. Maximum flow, 2.5 inches or 0.07 of an ounce, May 26. Total absorption, 759.1 inches or 21.09 ounces. Total flow, 3.5 inches or 0.097 of an ounce.

Ulmus Americana.—Gauge attached, April 11. Maximum absorption, 26.5 inches or 0.74 of an ounce, April 15. Maximum flow, 12.5 inches or 0.34 of an ounce, April 29. Total absorption, 155 inches or 4.30 ounces. Total flow, 256.8 inches or 7.13 ounces.

Pyrus Malus.—Gauge attached, April 11. Maximum absorption, 25.0 inches or 0.70 of an ounce, April 16. Maximum flow, 22 inches or 0.60 of an ounce, May 16. Total absorption, 175.3 inches or 4.85 ounces. Total flow, 290.7 inches or 8.07 ounces.

Water Gauges.

DATE	Acer saccharinum, detached root.	Pyrus Malus, detached root.	Castanea vesca, detached root.	Ulmus Americana, detached root.	Quercus alba, detached root.	Fraxinus Americana, detach'd root.
1874.						
Apr. 11,	–	—19.0	—7.5	—9.0	—45.0	—1.0
12,	–	—13.0	—2.0	—8.0	—15.0	–
13,	–	—7.5	—2.0	—9.0	—15.0	—3.0
14,	–	—8.5	—3.0	—11.5	—45.0	—2.0
15,	–	—21.0	—8.0	—26.5	—29.5	—5.0
16,	–	—25.0	—6.0	—25.5	—30.0	—4.0
17,	–	–	—6.0	—25.5	—30.0	—2.0
18,	–	–	—3.0	—18.5	—30.0	—1.0
19,	–	–	—5.5	—16.0	—30.0	—1.5
20,	–	–	—3.5	—5.5	—30.0	—1.5
21,	–	–	—2.5	0.3	—31.0	—0.5
22,	–	—16.0	—1.0	1.3	—31.0	—0.5
23,	–	—15.5	—0.3	2.0	—31.0	—1.0
24,	–	—11.0	–	1.8	–	–
25,	–	—9.5	—2.0	3.5	—32.0	—1.0
26,	–	—5.5	—2.0	1.3	—31.0	—1.0
27,	–	—4.5	–	5.0	—12.0	–
28,	–	—12.0	–	6.5	—28.0	—0.8
29,	–	—2.3	–	12.5	—32.0	—0.5
30,	–	—3.0	—0.5	10.0	—30.0	—0.8
May 1,	–	–	—0.3	8.0	—25.5	—0.3
2,	—14.2	—0.8	—1.5	7.5	—46.0	—0.5
3,	—17.5	—0.7	0.5	6.0	—20.0	—0.5
4,	—24.0	—0.7	—2.0	6.5	—15.5	—1.0
5,	—26.0	—0.3	—2.3	4.5	—17.0	—1.0
6,	—40.0	–	—3.0	10.0	—14.0	—1.0
7,	—38.0	0.3	—2.0	4.5	—12.0	—1.3
8,	—35.0	—0.5	—2.0	1.8	—10.0	—1.0
9,	—32.0	0.3	—1.0	4.0	—7.0	—1.0
10,	—69.0	0.8	—1.0	1.7	—1.0	—0.7
11,	—36.0	4.0	—1.3	3.5	—2.0	—1.3
12,	—19.0	3.5	—1.3	1.0	—3.0	—1.0
13,	—21.0	3.0	—0.5	2.0	—1.0	—1.0
14,	—21.0	6.5	—1.0	2.5	—2.0	—1.0
15,	—16.0	10.5	—1.0	2.3	—4.0	—1.3
16,	—9.0	22.0	–	3.0	—3.0	—1.0
17,	—3.0	11.5	—0.5	4.0	–	—0.5
18,	—4.0	16.5	—0.5	4.0	—1.5	—0.5
19,	—10.0	13.0	–	4.0	—1.0	—1.0
20,	—9.0	9.0	—0.5	4.0	—1.8	—0.6
21,	—7.0	6.0	—0.5	4.0	—1.0	—0.6
22,	—2.0	5.5	–	4.0	0.5	—0.3
23,	—5.0	5.0	—0.8	3.0	0.5	—0.5
24,	—3.0	4.0	—0.6	3.0	1.0	—0.5
25,	—5.0	4.0	—0.3	3.5	—1.0	—0.6
26,	—2.0	8.5	0.5	5.0	2.5	—0.5
27,	—7.0	5.0	—0.5	3.0	—1.0	—1.0
28,	—7.0	5.0	–	4.0	—1.0	—1.0
29,	—6.0	6.5	—0.5	4.0	—0.5	—2.0

Variation in Water Gauges—Concluded.

DATE.	Acer sacchari-num, detached root.	Pyrus Malus, de-tached root.	Castanea vesca, detached root.	Ulmus America-na, detached root.	Quercus alba, detached root.	Fraxinus Amer-icana, detach'd root.
1874.						
May 30,	—6.0	6.5	—0.3	4.0	—1.5	—2.0
31,	—5.0	7.0	—0 3	5.0	—1.5	—1.5
June 1,	—2.0	7.0	—0.3	5.0	—0.5	—1.0
2,	—5.0	8.0	—0.5	4.0	—1.3	—1.3
3,	—3.0	14.0	—0.3	4.0	—1.0	—1.3
4,	–	8.0	–	4.0	—2.0	—1.0
5,	–	9 0	–	4.0	—1.0	—1.0
6,	–	8.3	–	3.5	–	–
7,	–	12 0	–	4.0	–	–
8,	–	11.5	–	3.5	–	–
9,	–	12.0	–	3.0	–	–
10,	–	11.0	–	3.5	–	–
11,	–	10.5	–	3.0	–	–
12,	–	8.5	–	4.0	–	–
13,	–	6.5	–	2.5	–	–
14,	–	6.0	–	2.0	–	–
15,	–	5.0	–	3.0	–	–
16,	–	5.5	–	2.0	–	–
17,	–	5.0	–	2.0	–	–
18,	–	6.0	–	5.0	–	–
19,	–	5.0	–	2.0	–	–
20,	–	4.0	–	2.5	–	–
21,	–	6.0	–	1.5	–	–
22,	–	1.5	–	1.5	–	–
23,	–	4.0	–	1.3	–	–
24,	–	4.5	–	4.5	–	–
25,	–	2.0	–	1.5	–	–
26,	–	1.5	–	2.0	–	–
27,	–	3.0	–	2.0	–	–
28,	–	1.0	–	1.5	–	–
29,	–	0.5	–	1.0	–	–
30,	–	1.0	–	2.0	–	–

TABLE

Showing the fluctuations in Mercurial Gauges attached to the roots and trunks of trees, with descriptions of the specimens under observation. The figures denote inches of mercury, and when the minus sign is prefixed they indicate suction; otherwise, pressure. To convert inches of mercury into feet of water, multiply the number by 13.60, the specific gravity of mercury, and divide the product by 12. When the figures are omitted for one or more days after the record has begun, it shows that in consequence of some accident no observation could be made. The hour for the first observation of any day is seven A.M., that for the second, is noon, and that for the third, is six P.M.

NAME.	Height, feet.	Girth, feet, in.
Acer rubrum.	40	2 6
A. saccharinum.	60	6 4
Betula alba var. populifolia.	30	1 4
B. alba var. populifolia, root.	35	1 8
B. lenta, root.	65	3 10
B. lutea.	60	3 8
B. papyracea.	60	3 9
Ostrya Virginica.	45	2 10
Pyrus Malus, root.	35	2 9
Vitis æstivalis.	50	1 0
V. æstivalis, root.	40	10

Mercurial Gauges.

DATE.	Betula alba var. populifolia, detached root.	Vitis æstivalis, lower gauge.	Vitis æstivalis, upper gauge.	Vitis æstivalis, detached root.	Pyrus Malus, detached root.
1874—Apr. 12,	8.6	—	—	—	—
18,	22.5	—	—	—	—
21,	25.2	—	—	—	—
22,	26.0	—	—	—	—
23,	26.4	—	—	—	—
24,	27.0	—	—	—	—
25,	28.3	—	—	—	—
26,	27.3	—	—	—	—
27,	28.0	—	—	—	—
28,	28.3	—	—	—	—
29,	27.3	—	—	—	—
30,	28.3	—	—	—	—
May 1,	27.8	—	—	—	—
2,	28.2	—	—	—4.0	—

PHENOMENA OF PLANT-LIFE.

Fluctuations in Mercurial Gauges—Continued.

DATE.	Betula alba var. populifolia, detached root.	Vitis aestivalis, lower gauge.	Vitis aestivalis, upper gauge.	Vitis aestivalis, detached root.	Pyrus Malus, detached root.
1874—May 3, . . .	30.0	–	–	—3.7	–
4, . . .	30.0	–	–	—3.3	–
5, . . .	31.0	–	–	—3.3	–
6, . . .	31.0	–	–	—2.9	–
7, . . .	31.0	9.8	–	—2.1	–
8, . . .	31.0	12.4	–	—2.0	–
9, . . .	31.5	9.0	–	—1.3	–
10, . . .	28.5	14.0	–	—0.2	–
11, . . .	31 5	30.2	–	3.1	–
12, . . .	33.6	30.1	3.0	4.5	–
13, . . .	32.0	29.2	0.7	2.2	–
14, . . .	31.0	39.2	10.2	8.6	–
15, . . .	24.6	59.0	30.0	22.6	—1.4
16, . . .	25.0	58.0	35.0	17.4	2.5
17, . . .	26.9	50.0	30.0	8.8	4.8
	–	52.5	29.0	10.5	6.5
18, . . .	16.8	45 0	24.3	15.0	6.7
19, . . .	17.0	45.0	22.3	10.7	3.9
20, . . .	9.0	40.0	20.0	8.4	3.2
21, . . .	6.7	39.2	18.0	11 8	5.7
22, . . .	–	41 0	16.3	10.7	5 7
23, . . .	0.4	35.7	15.7	11.1	5.2
	—3.0	40.0	15.3	18.2	6.0
24, . . .	—2.0	45.4	14.4	19.5	6.0
	–	51.0	15 0	23.0	–
	–	53 9	15.0	–	–
25, . . .	—0 8	62.7	14.4	40.0	6.6
26, . . .	2.5	74.0	20.7	52.5	7.2
27, . . .	—3.0	66.0	23.5	42 9	–
	–	57.0	23.3	48.0	8.4
28, . . .	—5.3	49.2	23 3	47.0	7.0
	–	54.0	22.0	65.0	–
29, . . .	—5.4	57.3	22.2	60.5	7.2
	–	62.0	23 8	78.3	–
30, . . .	—5.0	59.8	24.5	66.5	9 0
	–	–	–	78.0	12 5
	–	–	–	72.0	6.2
31, . . .	—4.3	44.5	17.5	57.5	10 0
	–	50.2	18.6	63.0	13 3
	–	47.3	18.0	71.0	10.5
June 1, . . .	—3.0	48.0	17.0	54.3	9.0
	–	40.0	16.0	52.2	7.4
2, . . .	—5.0	34.0	15.0	37 0	6.2
	–	21.5	13.3	43.9	7.4
3, . . .	–	23.6	12.5	37.6	5.1
	–	19.5	11.3	42.1	7.1
4, . . .	—4.8	25.1	10.8	41.0	5.8
	–	.8	10.3	–	–
5, . . .	–	25.6	9.4	39.2	4.9
	–	23.8	9.1	43.6	7.0

PHENOMENA OF PLANT-LIFE.

Fluctuations in Mercurial Gauges—Continued.

DATE.	Betula alba var. populifolia, detached root.	Vitis æstivalis, lower gauge.	Vitis æstivalis, upper gauge.	Vitis æstivalis, detached root.	Pyrus Malus, detached root.
1874—June 6,	—	27.0	8 5	41.8	4.3
	—	26.8	7.9	48.0	5.7
7,	—6.0	30.0	7.5	42.0	4.8
	—	17.9	6.6	43.0	6.5
	—	—	—	44.8	5.6
8,	—	33.9	6.7	42.4	4.5
	—	6.9	4 4	41.4	6.8
	—	—	—	42.7	5.8
9,	—	10.0	3.5	35.8	3.8
	—	2.7	2.8	41.4	6.8
10,	—10.0	6.0	2.0	37.0	4.1
	—	—	—	38.3	7.3
	—	—	—	42.0	7.3
11,	—11.5	3.2	—0.3	34.2	4.0
	—	—	—	33.3	6.6
	—	—	—	32 2	6.3
12,	—9.2	11.8	1.1	31.5	4.1
	—	—	—	32.5	6.0
	—	—	—	32.9	6.0
13,	—10.8	0.2	—5.4	31.0	3.4
14,	—11.2	—4.0	—3.3	30.2	3.5
15,	—12.7	—8.0	—4.5	28.0	2.7
	—	—0.3	—5.0	35.0	5.8
16,	—13.0	1.1	—6.0	34.6	2.8
	—	—6.2	—6.6	38.6	4.5
17,	—12.0	3 8	—7.1	38.0	4.5
18,	—11.7	—3.4	—7.9	34.3	3.4
19,	—12.4	—8.7	—9.0	30.3	3.0
20,	—12.7	—6.3	—9.5	27.5	2.8
21,	—12.4	—8.7	—9.3	31.5	2.0
	—	—	—	35.6	2.5
22,	—12.2	—11.0	—9.1	34.3	1.3
	—	—	—	38.2	2.0
23,	—13.5	—11.0	—9.3	38.5	2.0
24,	—13.7	—10.6	—9.5	39.0	2.6
25,	—14.2	—11.7	—9.0	34.3	2.2
26,	—14 6	—8.5	—9.0	33.0	2.0
27,	—13.6	—8.7	—8.3	39.0	3.0
28,	—14.3	—11.3	—7.9	38.5	1.8
29,	—14.4	—12.7	—7.0	37.3	2.0
30,	—14.6	—11 0	—7.5	39.0	0.8
July 1,	—15.0	—	—	37.6	—
2,	—15.5	—	—	38.5	—
3,	—15.5	—	—	41.0	—
4,	—16.0	—	—	40.0	—
5,	—15.8	—	—	35.3	—
6,	—16.1	—8.6	—5.3	33.3	—
7,	—16.2	—	—	34.3	—
8,	—16.3	—	—	36.0	—
9,	—16.4	—	—	40.4	—

PHENOMENA OF PLANT-LIFE.

Fluctuations in Mercurial Gauges—Concluded.

DATE.	Betula alba var. populifolia, detached root.	Vitis aestivalis, lower gauge.	Vitis aestivalis, upper gauge.	Vitis aestivalis, detached root.	Pyrus Malus, detached root.
1874—July 10,	—16.5	—	—	44.0	—
12,	—15.8	—	—	37.5	—
13,	—15.0	—	—	32.7	—
14,	—16.6	—	—	33.5	—
15,	—16.4	—	—	37.4	—
16,	—15.8	—	—	42.0	—
17,	—16.6	—	—	48.0	—
18,	—16.6	—	—	40.4	—
19,	—16.7	—6.5	—5.0	41.6	—
20,	—16.7	—	—	42.2	—
21,	—16.5	—	—	41.0	—
22,	—16.6	—	—	36.0	—
23,	—16.6	—6.7	—4.4	34.0	—
24,	—16.8	—	—	38.5	—
25,	—16.8	—	—	39.7	—
26,	—17.0	—6.8	—5.0	42.0	—
27,	—17.0	—	—	37.9	—
28,	—17.4	—	—	36.5	—
29,	—17.4	—	—	35.4	—
30,	—17.2	—	—	35.6	—
31,	—17.9	—	—	28.4	—
Aug. 1,	—17.9	—	—	28.9	—
3,	—18.4	—	—	33.4	—
4,	—19.4	—	—	29.9	—
5,	—19.4	—3.0	—5.0	29.0	—
6,	—18.5	—	—	27.1	—
7,	—19.0	—	—	28.7	—
8,	—18.8	—	—	31.5	—
9,	—18.5	—	—	31.4	—
10,	—18.2	—	—	31.0	—
11,	—18.3	—	—	27.2	—
12,	—18.3	—	—	28.9	—
13,	—18.3	—	—	26.0	—
14,	—18.8	—	—	24.1	—
15,	—18.9	—	—	23.6	—
16,	—18.8	—	—	28.2	—
17,	—19.2	—3.5	—6.3	25.3	—
18,	—19.2	—	—	26.0	—
19,	—19.3	—	—	26.0	—
20,	—19.4	—	—	21.2	—
21,	—19.3	—	—	23.2	—
22,	—19.3	—	—	27.7	—
23,	—19.6	—	—	18.8	—
25,	—19.8	—	—	17.9	—
26,	—20.2	—	—	18.4	—
27,	—19.4	—	—	21.1	—
29,	—19.4	—	—	20.6	—
30,	—19.0	—	—	22.4	—
31,	—19.0	—	—	20.8	—
Sept. 14,	—	—	—	17.0	—

PHENOMENA OF PLANT-LIFE.

Fluctuations in Mercurial Gauges.

DATE.	Betula lutea, lower gauge.	Betula lutea, upper gauge.	Betula lenta, detached root.	Betula alba var. populifolia.	Ostrya Virginica.
1874—Apr. 9, . . .	14.2	–	–	9.5	–
	16.5	–	–	11.5	1.5
	17.0	–	–	11.4	0.9
10, . . .	18.5	–	–	13.2	1.6
	21.0	–	–	14.7	3.5
	20.6	–	–	13.7	0.5
11, . . .	22.2	–	–	15.4	3.8
	22.5	–	–	14.0	0.6
	21.2	–	–	0.3	—2.7
12, . . .	30.0	–	–	–	5.0
	29.6	–	–	0.3	5.2
	—0.1	–	–	0.2	—2.0
13, . . .	11.6	–	–	2.2	1.6
	—1.6	–	–	0.2	—1.5
	—5.7	–	–	—2.2	—1.4
14, . . .	9.9	–	–	4.0	1.3
	11.0	–	–	1.2	1.5
	11.5	–	–	0.3	1.5
15, . . .	22.6	–	–	0.5	2.6
	27.0	–	–	10.2	6.0
	27.0	–	–	10.5	6.3
16, . . .	27.7	–	–	11.8	7.2
	32.0	–	–	15.5	10.5
	31.6	–	–	11.4	10.2
17, . . .	34.0	–	–	14.8	11.4
	35.5	–	–	14.6	10.4
	36.1	–	–	–	9.6
18, . . .	50.5	–	–	29.0	16.5
	29.0	–	–	29.6	3.0
	34.4	–	–	27.0	0.7
19, . . .	38.0	–	–	10.3	4.8
	47.0	–	–	–	11.5
	45.0	–	–	–	7.5
20, . . .	49.0	–	–	–	8.2
	53.0	–	–	–	9.3
	53.6	–	–	–	9.0
21, . . .	57.6	–	–	–	8.9
	63.0	–	–	–	11.0
	52.2	–	–	18.0	5.0
22, . . .	29.6	–	–	10.3	2.5
	65.5	–	–	32.8	10.0
	–	–	–	26.0	7.3
23, . . .	–	–	–	24.4	5.4
	53.0	–	23.6	35.0	6.2
	52.6	–	26.0	33.6	2.0
24, . . .	35.0	5.0	63.0	20.8	3.4
	65.0	36.8	58.0	28.8	8.2
	52.6	19.8	52.0	25.0	5.8
25, . . .	48.8	22.0	51.4	32.2	3.3
	52.0	22.0	43.0	22.7	0.2

PHENOMENA OF PLANT-LIFE. 77

Fluctuations in Mercurial Gauges—Continued.

DATE.	Betula lutea, lower gauge.	Betula lutea, upper gauge.	Betula lenta, detached root.	Betula alba var. populifolia.	Ostrya Virginica.
1874—Apr. 25, . . .	43.0	15.5	30.0	20.0	—3.3
26, . . .	-	-	4.4	—0.6	2.2
	36.0	11.6	33.8	23.4	1.5
	33.0	8.7	33.0	15.4	—6.7
27, . . .	45.7	20.1	35.3	25.5	1.7
	58.0	34.2	37.6	24.3	7.0
	42.5	15.7	33.4	9.8	—6.7
28, . . .	46.3	—9.3	7.3	3.7	-
	36.6	12.3	19.2	—0.2	2.8
	26.0	3.2	20.0	2.2	—3.8
29, . . .	37.3	13.5	23.6	-	—0.2
	43.5	19.0	26.6	-	2.3
	44.8	20.0	30.0	-	1.6
30, . . .	41.6	17.7	-	7.3	2.6
	47.6	22.0	-	17.5	3.2
	50.0	23.0	-	6.8	5.0
May 1, . . .	42.6	16.5	-	-	0.2
	56.0	28.3	44.1	4.8	6.3
	54.0	26.6	46.0	8.5	6.2
2, . . .	42.0	15.0	55.5	6.5	0.4
	-	0.8	37.0	14.1	4.8
	-	0.8	29.0	8.8	4.0
3, . .	-	—5.5	32.0	—1.5	3.5
	-	18.5	40.0	30.0	11.0
	-	3.0	52.6	6.0	10.7
4, . . .	-	0.5	45.5	—1.4	10.0
	62.3	35.0	41.5	34.5	15.0
	38.6	13.0	52.4	13.2	15.1
5, . .	32.0	10.3	52.5	2.0	13.0
	54.0	29.4	51.2	2.7	16.6
	38.5	11.5	54.0	12.0	14.0
6, . .	14.0	-	50.7	0.6	10.2
	52.8	25.5	55.7	1.5	22.4
	27.0	0.2	61.5	1.2	20.8
7, . . .	4.3	—1.0	56.6	—4.1	22.5
	45.0	19.8	53.2	23.2	25.2
	36.2	11.0	57.3	10.4	25.2
8, . .	11.3	0.2	55.2	—0.6	23.6
	51.9	25.9	56.1	18.6	28.5
	31.2	4.4	59.2	7.4	22.1
9, . . .	6.4	-	59.1	—6.4	21.7
	63.2	36.4	60.3	32.0	31.6
	40.0	5.6	62.1	12.0	29.0
10, . . .	26.0	3.7	64.5	—4.9	33.4
	40.0	2.8	66.0	10.7	24.1
	2.6	-	68.0	—15.0	10.2
11, . .	6.6	1.0	56.0	—17.0	10.3
	39.0	20.0	56.9	7.4	10.5
	7.4	9.5	15.0	0.8	-
12, . . .	—1.6	8.1	61.0	10.1	-

Fluctuations in Mercurial Gauges—Continued.

DATE.	Betula lutea, lower gauge.	Betula lutea, upper gauge.	Betula lenta, detached root.	Betula alba var. populifolia.	Ostrya Virginica.
1874—May 12, . . .	11.0	—4.1	53.0	17.2	6.8
	—4.8	—3.6	54.6	—0 2	29.6
13, . . .	—3.5	—3.7	56.0	2.3	35.6
	-	-	-	-	—3.0
	5.6	—9.8	61.0	-	23.9
14, . . .	4.6	—2.9	58.0	—9.1	28 4
	23.0	—0.8	59.0	3.7	21.4
	—11.2	—2.6	58.7	—2.6	8.0
15, . . .	—13.1	—2.6	52.5	—11.0	15.7
	15.4	—2.3	54.1	—2.0	16.2
	—10.7	—2.4	53.0	—8 5	—3.6
16, . . .	7.8	—2.4	42.2	—0.3	9.0
	5.2	—2.0	44.2	4.3	—2.7
	—19.8	—2.0	43.0	—6.0	—17.1
17, . . .	—14.3	—2.1	38.0	—10.1	—10 5
	9.8	—2.1	37.0	—5 7	7.7
	10.0	—2.1	36.9	—4.4	7.9
18, . . .	2.7	—2.1	33.2	—4.0	—8.0
	3.0	—2.8	32.4	—2.7	—6.3
19, . . .	—2.7	—2.1	33.2	—4.0	—8.0
	—18.5	—2.2	32.8	—6.3	—21.7
20, . . .	—15.6	—2.2	29.4	—8.5	—16.0
	13.4	—2.1	32.2	—6.0	15.1
	16.3	—2.0	32.0	—10.1	13.0
21, . . .	—14.7	—2.1	30.2	—12.0	—9.3
	1.0	—2.2	30.6	—10.0	—6.0
	—1.6	—2.1	30.5	—9.8	—5.3
22, . . .	6.0	-	28.0	—8.4	—3.2
	—9.0	-	28.8	—7.3	—6.0
	—14.0	-	28 4	—10.6	—9.4
23, . . .	—14.0	-	25.0	—12.0	—7.1
	—14.2	-	28.3	—12.0	—5 8
	—14.5	-	28.3	—10.0	—5.0
24, . . .	—13.8	-	25.2	—15.0	5.0
	—12.8	-	29.5	—14.0	5.0
	—12.3	-	28.8	—14.0	4 8
25, . . .	—7 9	-	27.8	—14.0	—5.3
	—5.0	-	28.0	—13.0	—4.0
	—2.9	-	28.0	—11.9	—3.0
26 . . .	—1.0	-	25.5	—11.4	2.0
	—7.0	-	26.0	—11.7	—7.6
	—9.3	-	25.0	—10.4	6.3
27 . . .	—9.6	-	22.0	—11.1	—5.5
	—8.3	-	25.4	—8.6	—4.9
	—7.0	-	26.0	—8.5	—5.0
28, . . .	—7.4	-	22.6	—9.6	—5.0
	—6.0	-	25.9	—6.6	—4.4
	—6.0	-	25.6	—6.8	—5.0
29, . . .	—5.5	-	22.6	—5.5	—4.8
	—5.3	-	25.0	—4.0	—4.3

PHENOMENA OF PLANT-LIFE.

Fluctuations in Mercurial Gauges—Concluded.

DATE.	Betula lutea, lower gauge.	Betula lutea, upper gauge.	Betula lenta, detached root.	Betula alba, var. populifolia.	Ostrya Virginica.
1874—May 29,	—5.3	—	27.7	—4.9	—5.0
30,	—5.0	—	22.0	—6.0	—4.8
	—3.2	—	26.0	—4.0	—4.4
	—3.0	—	25.5	—4.4	—4.6
31,	—2.3	—	21.4	—5.3	—4.5
	—1.8	—	23.2	—4.4	—4.8
June, 1,	—1.9	—	20.0	—5.0	—4.5
	—1.5	—	21.0	—2.6	—5.0
	—2.0	—	20.4	—3.0	—4.8
2,	—2.5	—	18.0	—5.2	—4.7
	—1.2	—	16.4	—4.7	—4.6
3,	—1.2	—	16.4	—4.7	—4.6
4,	—2.1	—	17.3	—5.2	—4.6
5,	—2.2	—	17.0	—1.7	—4.6
6,	—2.0	—	17.0	—3.0	—4.6
7,	1.4	—	16.7	—3.4	—
8,	1.5	—	17.0	—2.4	—
9,	—0.4	—	14.7	—4.6	—
10,	-	—	14.2	—3.0	—
11,	0.3	—	10.0	—4.0	—
12,	0.3	—	10.0	—3.6	—
13,	0.1	—	5.0	—3.7	—
14,	—0.3	—	9.0	—4.0	—
15,	—0.4	—	10.0	—4.0	—
16,	0.4	—	11.5	—3.4	—
17,	1.0	—	12.0	—3.0	—
18,	1.8	—	14.3	—3.0	—
19,	1.0	—	12.6	—3.2	—
20,	1.4	—	12.0	-	—
22,	0.7	—	12.0	-	—
23,	1.0	—	12.4	-	—
24,	1.3	—	12.4	—0.7	—
25,	0.5	—	11.0	—1.0	—
26,	0.7	—	11.6	—0.6	—
27,	1.5	—	11.6	—1.0	—
28,	1.3	—	11.4	—1.4	—
29,	1.7	—	11.0	—1.0	—
30,	1.7	—	10.7	—2.0	—

Fluctuations in Mercurial Gauge.

Date.	Betula papyracea.	Date.	Betula papyracea.	Date.	Betula papyracea.
1874.		**1874.**		**1874.**	
May 3,	38.0	May 16,	—1.6	May 29,	—1.6
4,	36.0		—5.6		—2.8
	32.0	17,	—3.4	30,	—0.5
	39.0		—2.0		—1.8
5,	51.4		1.3		—2.8
	49.7	18,	—3.3	31,	4.5
	46.0		4.2		—3.6
6,	49.0	19,	—3.4	June 1,	4.6
	54.0		—2.4		—3.7
	48.0		—3.0		—2.8
7,	48.5	20,	—4.3	2,	—4.7
	51.2		—2.3	3,	—4.8
	45.8		—3.0	4,	—4.6
8,	32.4	21,	—3.3	5,	—4.4
	48.8		—2.4	6,	—4.4
	44.7		—1.3	7,	—4.5
9,	28.1	22,	—2.4	8,	—5.0
	51.5		—2.0	9,	—6.0
	47.0		—2.7	10,	—5.4
10,	39.0	23,	—3.0	11,	—6.5
	48.0		—1.1	12,	—6.0
	26.7		—1.0	13,	—6.5
11,	7.0	24,	—2.0	14,	—7.0
	33.3		0.6	15,	—5.4
	13.0		—1.8	16,	—5.0
12,	—2.7	25,	—2.4	17,	4.7
	35.9		—2.2	18,	—5.5
	20.0		—2.0	19,	—5.2
13,	—2.3	26,	—2.3	20,	—5.6
	17.5		—2.0	22,	—4.4
	—6.2		—2.6	23,	—4.4
14,	—2.5	27,	—4.0	24,	—5.0
	11.5		—0.6	25,	—6.0
	—2.8		—1.0	26,	—5.4
15,	—1.8	28,	—3.8	27,	—6.0
	7.1		—2.2	28,	—6.0
	11.6		—2.4	29,	—5.6
16,	5.4	29,	—3.8	30,	—6.0

PHENOMENA OF PLANT-LIFE. 81

Fluctuations in Mercurial Gauges.

ACER SACCHARINUM.

DATE.	Gauge 1.	Gauge 2.	Gauge 3.	Gauge 4.	Gauge 5.
1874—Mar. 21, . . .	18.0	18.0	—6.8	4.3	6.0
	19.0	19.3	—3.3	18.5	3.5
	—1.2	1.3	—3.5	5.0	—1.5
22, . . .	—4.0	—2.0	—3.5	3.0	—0.5
	—2.3	0.5	-	2.6	-
	—5.5	—6.0	—2.5	0.2	-
23, . . .	-	—5.3	—3.0	7.7	-
27, . . .	4.3	2.0	—4.0	—4.0	-
	22.0	24.0	2.0	15.2	8.2
	25.5	9.5	—2.0	6.4	—0.3
28, . . .	26.3	10.0	—3.0	6.7	-
	35.0	32.0	0.5	19.7	14.3
	16.0	17.0	1.0	15.0	—4.0
29, . . .	25.0	15.5	0.6	12.3	—4.0
	12.6	17.0	1.0	10.0	2.3
30, . . .	16.5	17.2	1.1	11.0	0.3
	26.4	24.0	1.0	14.0	24.0
	22.8	24.7	0.2	16.6	6.0
31, . . .	25.5	12.2	-	9.0	—23.0
	1.2	15.0	-	6.4	8.2
Apr. 1, . . .	3.3	10.2	1.0	2.7	9.0
	3.1	10.0	1.0	3.0	8.5
	18.0	16.3	3.0	11.8	16.9
2, . . .	19.0	16.0	1.5	13.2	12.5
	-	-	1.2	8.0	46.0
	-	-	0.6	14.6	8.0
3, . . .	-	-	3.0	5.0	0.2
	17.5	13.0	-	13.9	0.2
4, . . .	—2.2	—6.8	2.7	4.0	—2.0
	—1.7	—4.0	2 4	4.0	—1.4
	—1.0	-	2.2	3.6	—1.6
5, . . .	—1.6	-	1.0	3.8	—2.0
	—0.3	—2.2	1.0	1.4	-
	15.5	-	1.4	5.4	1.2
6, . . .	7.3	11.1	4.0	4.0	1 8
	25.1	6.0	1.4	0.5	1.2
	16.2	14.0	1.1	10.9	0.8
7, . . .	—11.7	3.0	1.1	1.0	-
	25 0	5.1	1.4	0.9	23.0
	21.0	12.3	1.0	11.1	4.1
8, . . .	—7.8	—1.5	1.1	1.8	—0.1
	26.1	6.1	1.1	11.3	13.1
	12.1	11.1	1.1	9.6	—0.2
9, . . .	—2.8	6.1	1.0	3.4	—1.7
	—1.0	5.2	1.0	2.9	1.1
	—4.0	4.0	1.0	2.3	1.4
10, . . .	—6.1	1.6	0.8	1.3	—1.2
	1.6	2.4	1.0	2.0	—1.1
	—4.7	2.5	1.0	1.4	—1.1

82   PHENOMENA OF PLANT-LIFE.

Fluctuations in Mercurial Gauges—Continued.

DATE.	Gauge 1.	Gauge 2.	Gauge 3.	Gauge 4.	Gauge 5.
1874—Apr. 11,	—16.0	—2.7	1.0	—2.3	3.1
	12.0	1.8	1.0	8.7	—2.8
	2.0	3.3	1.0	6.2	—1.1
12,	—0.7	—0.4	2.0	—0 3	—1.1
	-	-	1.0	-	-
	—10.7	—0.3	2.0	—0.9	—0.9
13,	—9.6	—0.3	2.2	4.0	3.0
	1.6	—0.5	1.0	—1.2	—3.0
	15.0	1.2	1.0	9.0	3.3
14,	—6.1	—0.4	1.0	—0.7	3.8
	26.2	3.1	1.0	11.4	13.6
	17.0	5.3	-	10.7	1.3
15,	5.9	5.6	1.0	8.4	—0.6
	7.9	6.4	0.9	8 2	—0.4
	0.5	6.1	1.0	6.9	—0.3
16,	—6.5	2.8	1.0	0.7	—0.3
	—0.5	3.0	0.9	0.7	—0.3
	—2.8	1.7	1.0	-	—0.3
17,	—5.3	1.3	1.0	—1.2	—0.2
	—3 6	1.1	1.2	—1.3	—0.2
	—3 3	1.0	1.1	—1.6	—0.3
18,	3.3	—2.4	1.5	3.4	2 2
	14.1	—0.7	1.0	11.0	21.0
	19.0	—1.0	0.9	12.0	4.7
19,	13.0	—1.4	1.0	8.4	—0.4
	15.0	—3.0	1.0	10.2	-
	14.4	—3.6	0.8	9.7	-
20,	5.0	2.0	0.9	2.0	-
	2.6	1.8	1.0	1.4	-
	2.3	1.9	1.0	1.0	-
21,	1.8	1.9	0.5	-	-
	2.1	2.0	0 8	2.2	2.0
	2.6	2.2	0.7	1.4	0.2
22,	—0.4	1.3	0.6	—4.5	-
	2.0	2.1	0.7	—5.3	0.3
	3.1	2.4	0.5	7.5	0.2
23,	1.1	1 8	0.5	1.0	0.1
	1.1	1.8	0.5	1.3	0.2
	1.0	1.8	0.4	1.0	0.2
24,	-	1.2	0.2	0.3	0.2
	1.2	2.0	—2.6	4.2	0.3
	1.0	2.0	—4.0	4.0	0.3
25,	—0.6	1.3	—5.3	-	0.2
	—0.6	1.4	—5.0	-	-
	—0.6	1.3	—5.0	—2.0	—0.2
26,	—2.1	1.0	—0.8	—3.5	1.0
	—1.7	1.2	—4.3	10.6	2.0
27,	—1.5	1.0	—4.5	8 0	—1.0
	—1.0	1.6	—4.2	8 7	0.4
28,	—3.4	0.6	—0.4	2.8	0.2
	—2.5	1.1	—8.3	8.8	0.3
	—2.0	1.3	—7.0	9.8	0.3
29,	—1.5	1.2	—5.6	5.5	0.5

PHENOMENA OF PLANT-LIFE.

Fluctuations in Mercurial Gauges—Concluded.

DATE.	Gauge 1.	Gauge 2.	Gauge 3.	Gauge 4.	Gauge 5.
1874—Apr. 30, . . .	—2.5	1.5	—5.9	1.8	0.7
	—2.6	1.5	—6.0	2.0	0.5
	—2.0	1.6	—5.8	2.5	0.6
May 1, . . .	—2.5	1.5	—5.6	1.1	0.4
	—2.6	1.9	—5.9	2.3	0.6
	—2.9	–	—5.9	–	0.6
2, . . .	—3.5	–	—5.8	–	0.5
	—3.7	–	—6.2	0.6	0.5
	—3.6	–	—6.2	0.2	0.6
3, . . .	—4.8	–	—7.0	—1.2	0.4
	—4.0	–	—6.6	3.0	0.7
4, . . .	—4.3	–	—7.0	—2.0	0.5
	—3.5	–	—6.0	2.1	0.5
5, . . .	—2.6	–	—3.3	—0.3	0.6
6, . . .	—2.5	–	—3.2	—2.1	0.5
	—1.0	–	—3.0	—0.8	0.8
7, . . .	—1.5	–	—3.0	—2.1	0.6
	—1.2	–	—2.7	—2.0	0.6
8, . . .	—1.1	–	—3.1	—2.1	0.5
	—0.6	–	—2.5	—1.2	0.9
9, . . .	—0.6	–	—2.8	—2.3	0.6
	—0.1	–	—2.1	—0.6	0.8
10, . . .	1.0	–	—2.8	—1.3	1.0
	1.3	–	—3.2	—1.3	0.9
11, . . .	0.3	–	—3.0	—1.3	0.8
12, . . .	1.2	–	—1.8	—2.6	0.1
	1.6	–	—0.5	—2.1	1.1
13, . . .	0.9	–	—1.2	—1.8	1.0
	2.0	–	—0.3	—2.0	1.3
14, . . .	0.1	–	—2.0	—2.0	1.0
	1.2	–	—2.4	—1.7	1.2
15, . . .	–	–	—4.2	—2.1	1.0
	1.4	–	–	—3.0	1.3
16, . . .	0.4	–	–	—2.0	1.1
	0.4	–	–	—1.5	1.1
17, . . .	0.3	–	–	—2.0	1.0
20, . . .	0.1	–	–	—2.1	1.0
21, . . .	0.6	–	–	—2.0	1.0
22, . . .	0.5	–	–	—1.8	0.8
23, . . .	0.3	–	–	—2.0	0.6
24, . . .	0.6	–	–	—2.0	1.2
25, . . .	1.0	–	–	—1.8	1.0
26, . . .	1.0	–	–	—1.3	1.4
27, . . .	0.4	–	–	—2.0	1.3
28, . . .	0.2	–	–	—2.0	1.4
29, . . .	0.2	–	–	—2.0	1.0
30, . . .	0.2	–	–	—2.0	1.0
31, . . .	0.9	–	–	—1.8	1.5
June 1, . . .	1.2	–	–	—1.9	1.4
2, . . .	0.5	–	–	—2.1	0.8

Fluctuations in Mercurial Gauges.

DATE	Acer rubrum.	Juglans cinerea.	DATE	Acer rubrum.	Juglans cinerea.
1874.			**1874.**		
Mar. 28,	2.1	–	Apr. 11,	3.9	1.8
	9.2	1.0		1.4	5.3
	2.5	1.5	12,	9.0	5.8
29,	3.2	1.5		6.6	2.5
	2.3	3.0	13,	7.8	7.2
30,	5.2	5.2		5.0	1.0
	3.2	6.2		10.0	6.2
	2.2	9.0	14,	11.4	10.8
31,	2.5	10.5		–	10.4
	2.4	2.8		13.1	7.3
Apr. 1,	2.0	5.0	15,	2.1	3.7
	2.0	4.0		4.2	4.6
	2.3	6.3		—1.0	2.4
2,	2.2	4.6	16,	—1.9	0.5
	2.1	2.6		—2.5	2.8
	2.0	6.8		—1.8	—0.4
3,	2.2	5.0	17,	—0.6	0.8
	2.4	4.0		—0.4	1.2
4,	2.2	1.8		0.3	1.1
	2.2	1.0	18,	1.7	1.1
	2.2	1.6		8.4	7.0
5,	2.2	1.0		2.6	6.6
	–	1.5	19,	—1.1	4.0
	–	0.9		3.4	5.8
6,	–	1.0		0.5	1.3
	–	1.5	20,	1.9	1.0
	–	2.7		2.0	1.0
7,	14.0	4.7		2.0	0.8
	9.8	3.7	21,	1.9	1.2
	11.1	5.0		2.0	1.0
8,	16.1	4.4		1.7	1.0
	16.4	4.0	22,	7.5	0.9
	9.3	5.1		3.0	–
9,	3.0	1.3		3.0	–
	3.9	2.0	23,	0.9	–
	1.8	0.9		1.2	–
10,	1.0	0.3		1.2	–
	4.1	2.0	24,	8.0	–
	—1.4	—0.7		6.0	–
11,	2.1	—0.5		2.0	–

PHENOMENA OF PLANT-LIFE.

TABLE—Showing the Temperature and amount of Cloudiness in Amherst during the months of March, April, May and June, 1874. By Prof. E. S. SNELL, LL.D., of Amherst College.

MARCH, 1874.

DAY OF MONTH.	TEMPERATURE.				CLOUDINESS.		
	7 A.M.	2 P.M.	9 P.M.	Mean.	7 A.M.	2 P.M.	9 P.M.
1,	20.4	40.6	35.0	32.0	8	-	1
2,	28.0	50.3	42.0	40.1	-	-	-
3,	33.3	54.0	49.5	45.6	1	4	8
4,	52.7	56.0	36.2	48.3	10	5	3
5,	26.9	40.4	30.0	32.4	-	-	-
6,	20.0	34.9	32.7	29.2	2	8	10
7,	29.0	33.9	33.7	32.2	10	10	10
8,	30.5	35.8	31.8	32.7	8	7	2
9,	29.2	29.5	23.0	27.2	8	7	5
10,	17.2	23.1	21.0	20.4	7	9	10
11,	20.3	19.1	17.0	18.8	10	10	10
12,	12.0	22.7	16.0	16.9	3	3	-
13,	13.0	20.0	17.5	16.8	6	4	-
14,	19.0	32.0	26.0	25.7	1	-	-
15,	21.3	40.1	33.5	31.6	-	-	-
16,	25.0	44.0	37.0	35.3	7	7	8
17,	34.7	40.5	39.9	33.4	10	10	10
18,	39.9	55.7	52.7	49.4	10	10	10
19,	45.9	54.0	51.8	50.6	10	10	10
20,	38.3	42.2	33.3	37.9	5	3	-
21,	28.0	50.0	39.0	39.0	-	7	-
22,	37.0	41.3	30.7	36.3	8	2	5
23,	26.0	29.7	19.5	25.1	1	8	-
24,	9.7	23.5	22.5	18.6	-	7	1
25,	22.0	46.7	37.7	35.5	-	2	-
26,	40.3	57.2	45.0	47.5	8	9	3
27,	28.1	39.0	29.2	32.1	1	1	-
28,	30.0	41.0	34.2	35.1	10	4	5
29,	26.0	33.3	25.5	28.3	9	-	-
30,	30.1	48.7	34.8	37.9	7	5	-
31,	30.0	32.2	27.2	29.8	9	10	7
Mean,	-	-	-	32.96	48 per cent. of sky.		

APRIL, 1874.

	7 A.M.	2 P.M.	9 P.M.	Mean.	7 A.M.	2 P.M.	9 P.M.
1,	18.7	34.0	31.3	28.0	-	-	10
2,	28.0	44.8	33.7	38.8	10	1	-
3,	33.7	42.0	35.0	36.9	9	3	4
4,	28.3	31.0	18.5	25.9	10	8	-
5,	21.5	41.0	28.2	30.2	-	2	10
6,	31.5	41.5	33.3	35.4	10	5	-
7,	32.4	42.3	36.3	37.0	10	10	-
8,	29.9	57.0	46.8	44.6	1	8	10
9,	37.0	44.0	40.0	40.3	10	10	10
10,	35.0	41.7	37.5	38.1	10	7	10
11,	34.0	45.2	31.0	36.7	-	9	-
12,	21.0	30.0	24.7	25.2	-	-	-
13,	23.3	42.3	33.5	36.4	-	-	-
14,	35.0	63.0	53.0	50.3	-	5	-
15,	51.2	61.0	54.0	55.4	9	9	-
16,	38.0	49.8	42.0	43.3	8	2	3
17,	37.0	33.7	31.5	34.1	10	10	10
18,	31.0	48.3	36.5	38.6	-	5	-
19,	35.3	61.0	48.0	48.1	10	-	2
20,	40.0	38.7	36.5	38.4	10	10	10
21,	38.0	49.0	38.0	41.7	10	5	9
22,	37.5	51.0	39.0	42.5	5	2	5
23,	37.6	39.0	34.0	36.9	10	10	10
24,	35.0	51.2	41.3	42.5	1	7	5
25,	37.2	34.7	32.0	34.6	10	10	10
26,	32.8	43.2	36.5	37.5	10	8	5
27,	39.0	48.0	35.0	40.7	6	1	-
28,	36.0	45.5	35.0	38.8	1	10	10
29,	33.9	36·3	34.3	34.8	10	10	9
30,	33.0	41.0	39.8	37.9	9	8	2
Mean,	-	-	-	.38.32	56 per cent. of sky.		

PHENOMENA OF PLANT-LIFE.

TABLE—Showing Temperature, etc.—Con.

MAY, 1874.

Day of Month.	Temperature.				Cloudiness.		
	A.M.	2 P.M.	9 P.M.	Mean.	7 A.M.	2 P.M.	9 P.M.
1,	39.5	46.0	43.0	42.8	1	8	9
2,	44.0	43.0	36.5	41.2	8	9	-
3,	43.8	57.0	46.5	49.1	-	-	-
4,	40.9	63.5	51.0	51.8	3	9	5
5,	46.0	58.9	48.3	51.1	7	8	2
6,	43.0	59.0	47.0	49.7	-	8	10
7,	39.0	49.3	37.0	41.8	-	8	-
8,	41.0	58.0	45.0	48.0	7	7	-
9,	44.5	68.6	62.5	58.5	8	5	-
10,	71.0	85.0	54.8	70.3	5	-	2
11,	49.8	60.8	44.2	51.6	9	2	-
12,	44.0	66.0	50.0	53.3	-	-	-
13,	50.2	80.5	63.0	64.6	2	2	-
14,	62.0	78.5	60.0	66.8	1	1	-
15,	54.0	74.0	56.0	61.3	-	1	3
16,	50.2	53.0	51.5	51.6	10	10	10
17,	52.8	68.0	55.0	58.6	7	-	1
18,	52.4	58.0	54.0	54.8	8	10	10
19,	51.5	61.1	49.0	53.9	5	3	-
20,	48.3	64.0	56.5	56.3	3	9	10
21,	51.5	57.1	49.5	52.7	10	10	10
22,	48.0	57.0	49.0	51.3	8	8	6
23,	50.0	66.8	56.0	57.6	-	8	1
24,	57.2	71.7	59.5	62.8	7	8	9
25,	57.3	59.8	60.3	59.1	10	10	9
26,	62.0	61.0	50.2	57.7	2	9	-
27,	53.8	72.0	60.5	62.1	1	3	1
28,	58.2	80.0	64.0	67.4	-	-	-
29,	62.0	86.0	67.0	71.7	-	1	6
30,	62.2	82.6	69.0	71.3	2	1	7
31,	63.0	82.8	68.3	71.4	10	7	10
Mean,	-	-	-	56.52	45 per cent. of sky.		

JUNE, 1874.

1,	62.5	66.8	55.0	61.4	6	8	9
2,	51.3	65.0	55.5	57.3	-	1	-
3,	54.5	68.3	55.0	59.3	8	5	10
4,	57.3	66.7	64.5	62.8	10	10	9
5,	62.2	71.0	60.8	64.7	10	10	10
6,	67.0	75.5	68.0	70.2	10	10	1
7,	67.0	77.5	71.8	72.1	10	9	10
8,	71.0	78.0	66.2	71.7	-	3	3
9,	63.8	81.0	66.5	70.4	7	3	-
10,	69.0	77.0	62.0	69.4	7	1	-
11,	57.0	62.0	56.0	58.3	8	10	10
12,	56.5	68.0	64.0	62.8	10	9	8
13,	56.0	61.0	53.5	56.8	5	7	-
14,	55.5	72.9	61.0	63.1	-	7	-
15,	59.0	77.5	64.5	67.0	1	5	1
16,	65.0	74.0	65.0	68.0	6	7	10
17,	64.5	71.7	63.2	66.5	10	10	8
18,	64.7	71.8	63.7	66.7	10	8	7
19,	62.9	69.0	54.0	62.0	2	8	10
20,	54.2	65.2	59.5	59.6	8	9	9
21,	58.5	75.0	66.8	66.8	10	6	8
22,	60.0	82.3	71.5	71.3	5	3	5
23,	70.0	85.0	71.0	75.3	4	8	5
24,	66.4	72.1	62.0	66.8	1	-	5
25,	61.0	75.5	67.0	67.8	5	4	8
26,	64.0	70.0	60.1	64.7	10	9	10
27,	63.0	77.2	68.7	69.6	10	1	-
28,	64.5	87.0	76.0	75.8	-	1	-
29,	73.3	93.0	71.0	79.1	7	2	9
30,	68.2	76.5	60.0	68.2	2	5	-
Mean,	-	-	-	66.18	58 per cent. of sky.		

TABLE

Showing the Percentage of Water in the wood and bark of the branches and roots of certain species of trees at different seasons of the year.

GENUS.	Species.	Description.	PERCENTAGE OF WATER.			
			Feb.	April.	Sept.	Dec.
Abies	Canadensis.	One year,	48.66	–	–	–
		Two year,	49.62	–	–	–
		Root,	–	55.96	–	–
		Dead twig,[2].	18.76	–	–	–
Abies	excelsa.	One year,	45.50	–	–	–
		Two year,	44.28	–	–	–
		Dead,[2].	17.03	–	–	–
Acer	rubrum.	One year,	44.88	–	–	–
		Two year,	44.71	–	–	–
Acer	saccharinum.	One year,	46.50	–	48.10	47.36
		Two year,	47.13	–	44.05	47.00
		Sap-wood,	–	–	–	41.23
		Heart-wood,	–	–	–	40.12
		Dead,	18.85	–	–	–
		Root,	–	41.44	44.05	–
Æsculus	Hippocastanum.	One year,	49.14	–	59.68	–
		Two year,	46.08	–	59.05	–
Ailantus	glandulosa.	One year,	48.56	–	–	–
		Two year,	46.00	–	–	–
Alnus	incana.	One year,	50.47	–	–	–
		Two year,	51.45	–	–	–
Betula	alba v. populifolia.	One year,	46.24	54.97	53.90	–
		Two year,	42.00	55.64	48.52	–
		Root,	–	–	42.63	–
		Dead,	15.13	–	–	–
Betula	lenta.	One year,	38.25	–	–	41.80
		Two year,	40.54	–	–	40.73
		Root,	–	–	49.61	–
		Dead,	13.65	–	–	–
Carpinus	Americana.	One year,	38.70	–	57.68	–
		Two year,	39.41	–	48.69	–
		Dead,	13.84	–	–	–
Carya	amara.	One year,	–	–	–	33.26
		Two year,	–	–	–	31.23
		Root,	–	54.32	–	–
Fagus	ferruginea.	One year,	44.42	–	–	–
		Two year,	44.69	–	–	–
Juglans	cinerea.	One year,	45.51	–	54.22	–
		Two year,	46.73	–	51.41	–
Nyssa	multiflora.	One year,	50.95	–	51.14	–
		Two year,	48.93	–	50.93	–
Pinus	Strobus.	One year,	–	56.31	62.90	–
		Two year,	–	55.52	58.34	–
		Dead,[2].	11.90	–	–	–
		Root,	–	–	67.65	–
Platanus	occidentalis.	One year,	54.46	52.55	–	–
		Two year,	51.44	53.79	–	–
Populus	tremuloides.	One year,	49.77	–	53.30	–
		Two year,	50.86	–	51.00	–
Prunus	Persica.	One year,	46.13	–	–	–
		Two year,	40.39	–	–	–

Percentage of Water in Trees—Continued.

GENUS.	Species.	Description.	PERCENTAGE OF WATER.			
			Feb.	April.	Sept.	Dec.
Prunus	serotina.	One year,	–	–	50.00	–
		Two year,	–	–	50.34	–.
		Dead,	17.37	–	–	–
Pyrus	communis.	One year,	49.85	55.39	54.05	–
		Two year,	47.70	54.03	51.48	–
		Root,	–	–	60.39	–
Pyrus	Malus.	One year,	49.49	48.98	56.18	–
		Two year,	44.75	46.76	54.49	–
		Root,	–	64.82	54.78	–
		Dead,	12.88	–	–	–
Quercus	alba.	One year,	38.01	41.24	43.06	–
		Two year,	35.23	36.74	39.51	–
		Root,	–	53.07	51.28	–
		Dead,	15.47	–	–	–
Salix	alba.	One year,	49.88	–	53.07	–
		Two year,	51.65	–	49.73	–
		Root,	–	–	68.38	–
Tilia	Americana.	One year,	55.10	–	48.62	–
		Two year,	53.93	–	55.97	–
Ulmus	Americana.	One year,	41.37	–	57.14	–
		Two year,	39.77	–	52.31	–
		Root,	–	45.26	43.19	–
		Dead,	13.46	–	–	–
Vitis	æstivalis.	One year,	41.86	43.77	–	–
		Two year,	41.08	43.66	–	–
		Root,	–	55.11	–	–

TABLE

Showing the specific gravity of the Sap collected from various trees in Spring, with observations concerning the cane sugar, glucose and starch contained in them. By CHARLES WELLINGTON, B. S.

Number.	GENUS.	SPECIES.	Date—1874.	Specific gravity at fifteen degrees Centigrade.	PERCENTAGE COMPOSITION. Sugar.* Grape.	PERCENTAGE COMPOSITION. Sugar.* Cane.	PERCENTAGE COMPOSITION. Starch.	TROMMER'S COPPER REDUCTION TEST. Quantity of sap required to reduce ten cubic centimeters of Fehling's solution. Fresh Sap.	TROMMER'S COPPER REDUCTION TEST. Sap after being treated with hydrochloric acid.
1	Vitis	æstivalis.	May 15.	1.002	0.000	0.000		—	—
2	V.	æstivalis.	May 5.	Gum	0.000	0.000	Did not succeed in any instance in obtaining a reaction with iodine.	—	—
3	Acer	saccharinum.	Mar. 26, 28.	1.015	trace	2.777		1,360.0 cubic cent.	1.8 cubic cent.
4	A.	rubrum	Mar. 27, 28.	1.010	0.012	1.458		410.0 "	3.4 "
5	A.	rubrum.	Apr. 8.	1.010	trace	0.833		500 "	6.0 "
6	A.	rubrum.	Apr. 8.	1.007	trace	0.769		500. "	6.5 "
7	A.	Pennsylvanicum.	Mar. 30, 31.	1.010	trace	1.428		1,300. "	3.5 "
8	Pyrus	Malus.	May 14.	†	0.000	0.000		—	—
9	Cornus	alternifolia.	May 21.	1.007	trace	0.000		—	—
10	Platanus	occidentalis.	Apr. 6.	1.005	0.892	0.000		5.6 "	6.0 "
11	Juglans	cinerea.	Mar. 26.	1.010	0.104	1.284		48.0 "	3.6 "
12	J.	nigra.	Apr. 21, 27.	1.010	0.139	1.249		36.0 "	3.6 "
13	Ostrya	Virginica.	Apr. 25, May 2.	1.002	0.303	0.000		16.5 "	16.0 "
14	Betula	lutea.	May 25, 2.	1.005	0.625	0.000		8.0 "	8.0 "
15	B.	lutea.	May 20.	†	0.000	0.000		—	—

* Taking it for granted that the reduction of the copper solution in the several instances, was due entirely to the presence of glucose, the several percentages of sugar, as given above, are correct. This, however, for obvious reasons, remains an assumption, to some extent.

† The quantity of sap was insufficient to allow of taking the specific gravity.

REMARKS.

Specimen No. 2 was the colorless, translucent gum which exudes freely from the wood of the root and stem of the grape vine, at any time during the long period of nearly eight months, when the vital force is dormant. It was entirely free from grape sugar, cane sugar, or starch. When treated with water, it swelled up and appeared to be partly soluble and partly not. The large amount of ash contained an abundance of lime.

Specimen No. 5 was sap from a red maple which had been girdled about two years previous. No. 6 was sap from a red maple, in a normal condition, which stood not far from No. 5. It was placed in the list in order that it might be compared with No. 5.

Specimen No. 8 was a very small quantity of sap from an apple tree. When brought in, it very much resembled cider in color. It had an unpleasant, sour taste.

Specimen No. 13 was sap from an ironwood. Though somewhat turbid, this sap contained no solid particles which could be separated by filtration. About two quarts of the sap which flowed on the day of May 7th, and the same amount which flowed during the following night, were collected and allowed to stand in the laboratory for some months. They became milky in a very short time, and fermented quite rapidly, emitting a very offensive odor. There was no difference between the two in this respect, so far as could be determined by their external appearance.

On the seventh of May, the sweet exudation from the hickory was tested for cane sugar. By means of alcohol it was removed to a glass plate, and when dry was examined under the microscope; it was also treated with Fehling's copper solution; but neither test showed a trace of cane sugar. Grape sugar was indicated to be present in abundance.

Gas from Sap of Acer saccharinum.

On the twenty-seventh of April, two and a half quarts of the first run of sap from a sugar maple was collected for examination in regard to the composition of the gas contained in it. By boiling, gas was obtained from this sap which measured 31.2 cubic centimetres at 18° C. By introducing a certain amount of potassium hydrate, the volume was reduced to 29.5 cubic centimetres at 18° C., owing to the absorption of carbonic acid by the potassium hydrate. By inserting a certain amount of gallic acid, the volume was again reduced to 22.5 cubic centimetres at 18° C., due to the absorption of oxygen, thus leaving 22.5 cubic centimetres of nitrogen.

Composition by Volume.

	Gas from Sugar Maple.	Atmospheric Air.
Nitrogen,	72.213	79.02
Oxygen,	22.435	20.94
Carbonic acid,	5.352	0.04
	100.000	100.00

PHENOMENA OF PLANT-LIFE. 91

Large quantities of sap from specimens of Vitis æstivalis, Acer saccharinum, Acer rubrum, Juglans nigra, Ostrya Virginica, and Betula lutea, have been evaporated preparatory to making analyses of their mineral constituents. This work has not yet been accomplished, for lack of time.

Explanation of Figures.

Fig. 1 represents two nodes of the squash vine.
    A is the petiole of a leaf showing vertical striæ.
    B, a staminate flower on a long peduncle.
    C, a branching tendril exhibiting the mode of attachment to a support, and the double reversed spiral of the portion between the support and the base of the tendril, by which all the branches of a tendril are made to bear their share of the strain, if they secure an attachment; and by which also great elasticity is given to the tendril, and the liability of rupture largely diminished.
    D, nodal roots.
    E, a pistillate flower with a short peduncle.
    F, a lateral branch of the vine.
    G, a tendril which, having failed in finding a support, has coiled upon itself and turned back towards the older portion of the vine.

Fig. 2 illustrates the structure of the tip of a squash rootlet, the cells of the epidermis being often produced into root-hairs consisting of single elongated cells, which increase immensely the absorbing surface.

Fig. 3 shows a transverse section of a rootlet.
    A, epidermis with root-hairs.
    B, ordinary cellular tissue.
    C, a fibro-vascular bundle.
    D, loose parenchyma of the central portion of the rootlet.

Fig. 4 is a longitudinal section of rootlet.
    A, epidermis with root-hairs.
    B, cellular tissue.
    C, a dotted duct.

Fig. 5 illustrates the structure of cork or periderm from a squash. The cells are large, thin-walled, dry and brown. They are developed in a radial manner from any highly vitalized cellular tissue, when it is exposed to the air. Every place upon the soft parts of a growing plant which is wounded soon covers itself with this protecting layer of cork.

Fig. 6 is a transverse section of a squash vine.
    A, the irregular internal cavity.
    B, fibro-vascular bundles.
    C, the outer green layer of the bark.

Fig. 7 is a transverse section of the petiole of a leaf.
    A, internal cavity.
    B, fibro-vascular bundles.
    C, vertical dark green striæ between the bundles, consisting of parenchyma containing chlorophyl.

Fig. 8 exhibits a transverse section of the branch of a tendril.
    A, the inner sensitive surface of loose cellular tissue, which contracts and expands as the branch coils and uncoils.
    B, bast fibre or elongated fusiform cells.
    C, fibro-vascular bundles

Fig. 9 represents the andrœcium of the staminate flower with connected sinuous anther cells, which are open and discharging pollen grains.

Fig. 10 is a pollen grain of spherical form and covered with projecting spines.
    A is the opening in the outer membrane through which the tube develops after its lodgment on the stigmatic surface of the pistil.

Fig. 11 shows the gynæcium of the pistillate flower.
    A, ovary.
    B, style.
    C, stigma.

Fig. 12 is a vertical section of the pistil.
    A, the receptacle, or stem.
    B, the wall of the ovary, the fibres of which are arranged in three distinct layers. The outer and inner ones have the fibres extending from the base to the apex of the ovary, or young squash, while the central one consists of fibres running around the ovary at right angles to the other two.
    C, ovules imbedded in loose cellular tissue.
    D, canal of the style through which the pollen tubes find their way to the ovules.

Fig. 13 represents a transverse section of the ovary, showing the three layers of the tissues of the wall and the cells of the ovary with ovules attached to the inner edges of the carpellary leaves.

Fig 14 exhibits the propagating pit with the squash in harness, and the squash root of a second vine attached to a mercurial gauge to show the pressure of the sap.
    A, the box in which the squash was placed.
    B, the lever to support the weights.
    C, the root from which the principal vine grew.
    D, the root of the vine which was cut off when eight weeks old, and connected with a gauge.
    E, mercurial gauge.
    F, scale to indicate the variations in the position of the lever.

Fig. 15 gives a view of the apex and lower side of the squash, after it had completed its growth, and been taken from the box in which it had been confined.

PHENOMENA OF PLANT-LIFE.

FIG. 16 shows the top of the squash, with the marks of the harness irons upon it.

FIG. 17 represents a piece of the root of an apple tree which penetrated a bed of coarse, dry gravel, to the depth of more than eight feet, and as it enlarged adapted itself to the spaces between the pebbles, and in some cases entirely inclosed them.

FIG. 18 illustrates the manner in which the roots of a black spruce grew on Moose Mountain, in New Hampshire. The soil was only a few inches deep, and below was solid rock, so that as the horizontal roots increased in diameter, they lifted themselves out of the ground, and of course raised the entire tree every year.

FIG. 19 shows how the heart of a yellow birch, growing on a ledge in Hanover, N. H., has been carried upward and outward by the annual deposition of wood, from the rock on which it must have rested when the seed germinated. The peculiar thickening of the trunk and roots near the base is often seen in trees on exposed situations.

FIG. 20 is a section of the stem of a tree *(Hibiscus splendens)* about four inches in circumference, from which all the bark and most of the wood was removed. A portion of the outer layer of sap-wood, one inch long and seven-sixteenths of an inch in circumference, was left to convey the sap to the foliage, which had a surface of twenty-five hundred square inches. Not a leaf wilted, but the supply of water was abundant for the growing tree.

FIG. 21 exhibits a section of a similar stem from a portion of which the wood was entirely removed, while the greater part of the thick, succulent bark remained. The foliage had a surface of five hundred square inches, while the amount of living bark which formed the connection between it and the roots was at least five times as large as the piece of sap-wood in the preceding figure. The leaves wilted as quickly and completely as if the stem had been entirely severed.

FIG. 22 is a piece of wood from a red maple, which threw out a callous from its ends like a grape cutting, and grew, although it had neither roots nor buds.

FIG. 23 shows a section of an elm root which was girdled, inclosed in a glass tube so as to exclude the air, and then replanted in the earth, its connection with the tree remaining intact. A new bark and layer of wood formed from the cambium which had been previously deposited.

FIG. 24 exhibits a section of the trunk of a small elm, upon the bark of which a horizontal incision was made, and above this four vertical incisions three inches long. The four quarters of the bark were then turned up, and a piece of tinned copper, one inch wide, was wrapped around the wood. The bark was then replaced, covered with waxed cloth, and securely fastened down. This was done on the thirtieth of May.

Fig. 25 shows the section as arranged for the experiment of determining whether the new layer of wood would be developed from the old wood or from the bark.

Fig. 26 represents the appearance of the new wood (b) which was deposited upon the metal, (a) after the removal of the bark in September.

Fig. 27 gives the microscopic structure of a horizontal section of the elm wood and bark directly over the metal. Next to the tin was a thin layer of parenchyma, (a) connected to the inner layer of bark by medullary rays, (c) which were as numerous as in the other parts of the new wood, and passed directly from the bark to the metal, whether examined in a horizontal or vertical section. The cork cells, (f) bast (d) and parenchyma (e) of the bark, and the woody fibre, (b) ducts (g) and medullary rays of the stem, are clearly visible in this section.

Fig. 28 is a view of the longitudinal section of the branch of an apple tree which was girdled in May, 1870. After growing four years and bearing fruit, it was cut in 1874. There were then many dead twigs upon it, and it was evidently in declining health. The section shows how the sap-wood was becoming dry, and changing into heart wood, so that the channel for the transmission of the sap from the roots to the leaves was almost closed. The girdling was complete, so that the elaborated sap from the leaves could not descend below it.

A is the top of the nearly horizontal branch.

Fig. 29 shows how a branch of a wild grape vine, after being girdled, formed new wood from both above and below, and thus made a new passage for the downward flow of the sap. The wood developed from beneath the girdle was formed from sap elaborated in other branches.

Fig. 30 is a section of the stem of a young bass tree, which shows that when there is no foliage below a place girdled early in the season, there can be no deposition of new wood, while it may be as abundant as usual above the girdle.

Fig. 31 represents a section of a stem of a young red maple, girdled June twenty-third, which is enlarged above the girdle, but not below.

Fig. 32 exhibits a similar section, girdled July twenty-first, upon which was produced a new growth of both wood and bark, which resulted from the fact that the cambium layer was so far organized by midsummer as to furnish a conducting medium for the elaborated sap.

Fig. 33 shows the microscopic structure of the ordinary bark of a young red maple.

A, periderm or cork.
B, primary parenchyma.
C, secondary parenchyma.
D, bast fibres.
E, woody fibre of trunk.
F, vessels or ducts in wood.
G, medullary rays connecting bark and wood.
H, recent parenchyma of inner bark.

PHENOMENA OF PLANT-LIFE. 95

Fig. 34 represents the same elements of the new bark formed on the place girdled, July twenty-first, the periderm being of a reddish brown color.

Fig. 35 is a view of the section of a weeping willow tree, to illustrate the mode of growth in trees from which the bark has been loosened by freezing.

A is sap-wood formed on the inside of the bark and disconnected from the wood of the trunk.

B is new wood and periderm formed on the old wood, to which a portion of the cambium cells remained attached when the old bark was torn off by the frost.

C, roots developed from the uninjured stem under the old disrupted bark, and extending to the earth, a distance of more than fifteen inches.

Fig. 36 exhibits a specimen of a pendant weeping willow branch, which was girdled in June last. The growth was on the lower side of the girdled place, showing that the flow of the elaborated sap is not necessarily downward, but *root*-ward.

Fig. 37 is a view of a pistillate plant of mistletoe, with evergreen coriaceous leaves and white berries, growing on the limb of an oak.

A represents the parasitic roots of the mistletoe in the sap-wood of the oak. As the oak was dead beyond the large cluster of the parasite, it seems that it was injured by the loss of its sap.

Fig. 38 illustrates the natural grafting of two trunks of white pine.

A is the smaller trunk, a branch of which is seen to grow through the wood of the larger one. The union of wood is perfect, and the elaborated sap from B has flowed so freely over the connecting branch, that A is larger below, and B larger above the place of junction.

C is the knot in the heart of B, formed of the base of the limb, in the axil of which D, the connecting branch, became fastened in the beginning of the operation.

Fig. 39 shows the grafted roots of a white pine stump, the points of union being very numerous.

Fig. 40 exhibits a section of a small white birch tree, one of whose branches has become grafted to it in consequence of being caught in the axil of a branch above.

Fig. 41 represents a section of the trunk of a small aspen, around which a vine of bitter-sweet has twined so closely as to prevent the root-ward flow of elaborated sap. The growth therefore follows the bitter-sweet in a spiral direction.

Fig. 42 shows a longitudinal section of the preceding specimen. The wood is seen to have formed from above so as to cover the vine, while immediately below it there has been no growth whatever.

Fig. 43 exhibits the dead wood of an apple tree limb, which was deposited so that the fibres run in a spiral direction.

Fig. 44 represents the variations of pressure, as indicated by the mercurial gauges, on the twenty-first of April, 1873, observations having been taken every hour, from twelve A. M., to twelve P. M. Every vertical line marks an hour, and every horizontal line an inch on the column of mercury. Zero represents the point where there is neither pressure outward from the tree, nor suction inward.

The line A shows the record of the sugar maple, which at midnight exhibited a suction equal to −6 inches, and at 7 A. M. had increased this to −22.9 inches. As soon as the sun warmed the tree, the mercury began to rise and at 9.15 A. M., had reached 16.3 inches. Then it declined very gradually till at 12 P. M. it was at −3 inches. The temperature at 7 A. M. was 37° F.; at 2 P. M. was 50.1° F.; and at 9 P. M. it was 39.5° F.

The line C marks the fluctuations of the mercury in the lower gauge of the black birch, which was at the level of the ground, and the line B shows the pressure in the upper gauge, which was placed 30.2 feet above the lower one. The remarkable fall, indicated as occurring at 12.45 P. M., was caused by boring into the tree near the ground for the purpose of determining whether the tree was acting simply as a cylinder of water filled by a force from beneath, as seemed evident from the correspondence between the two gauges. The reduction and restoration of pressure from simply opening and closing the orifice were so rapid and extraordinary as to lead to the conclusion that the force operating to produce the pressure was simply the absorbent power of the roots, and this led to the application of a gauge directly to a root, with the surprising result described on page 253.

The proportion borne by the cuts to the natural size of the object represented is expressed by a fraction under the figure. Thus, in figure one, the fraction $\frac{1}{6}$ indicates one-sixth the natural size, while in figure three, the fraction $\frac{50}{1}$ indicates that the object is magnified fifty times.

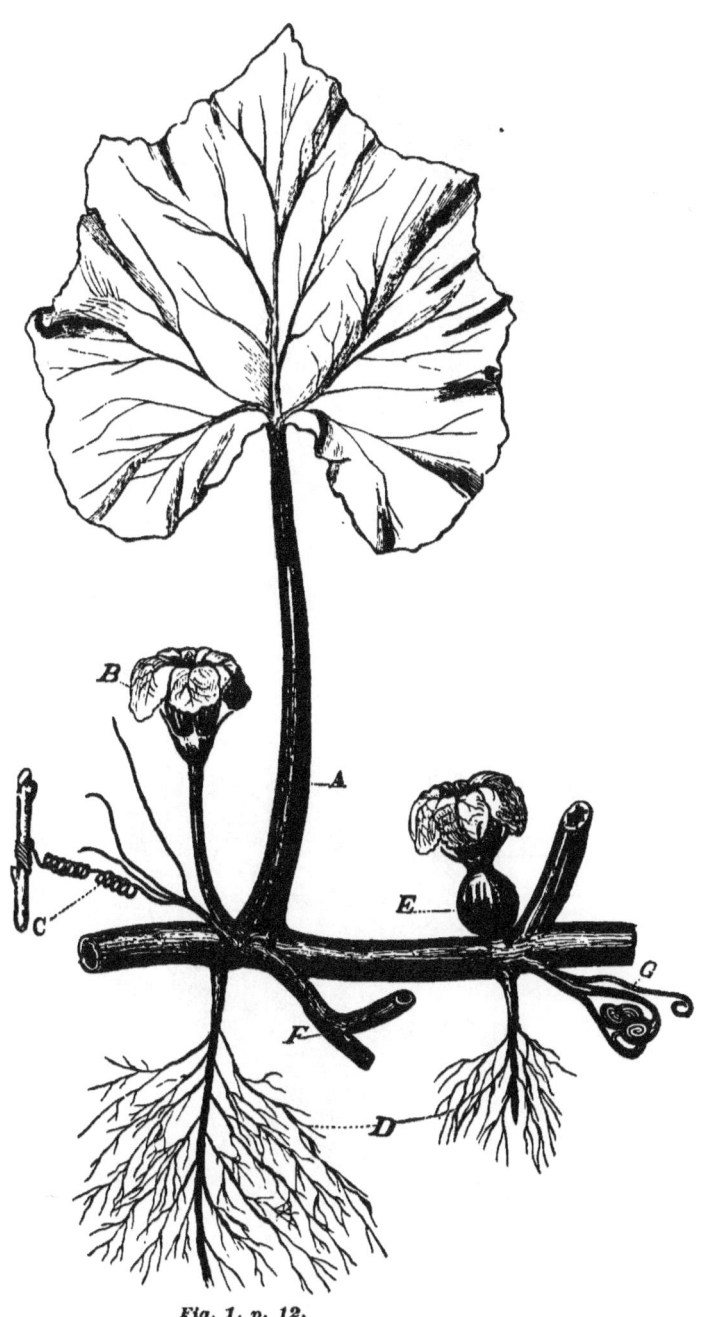

Fig. 1, p. 12.
⅙

98   PHENOMENA OF PLANT-LIFE.

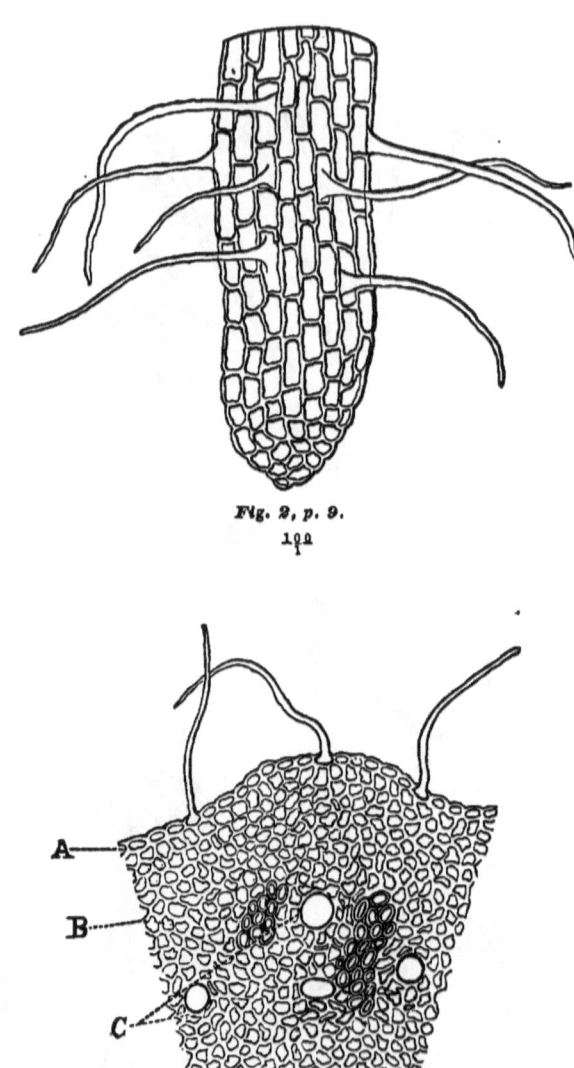

Fig. 2, p. 9.
¹⁰⁰⁄₁

Fig. 3, p. 9.
⁵⁰⁄₁

PHENOMENA OF PLANT-LIFE. 99

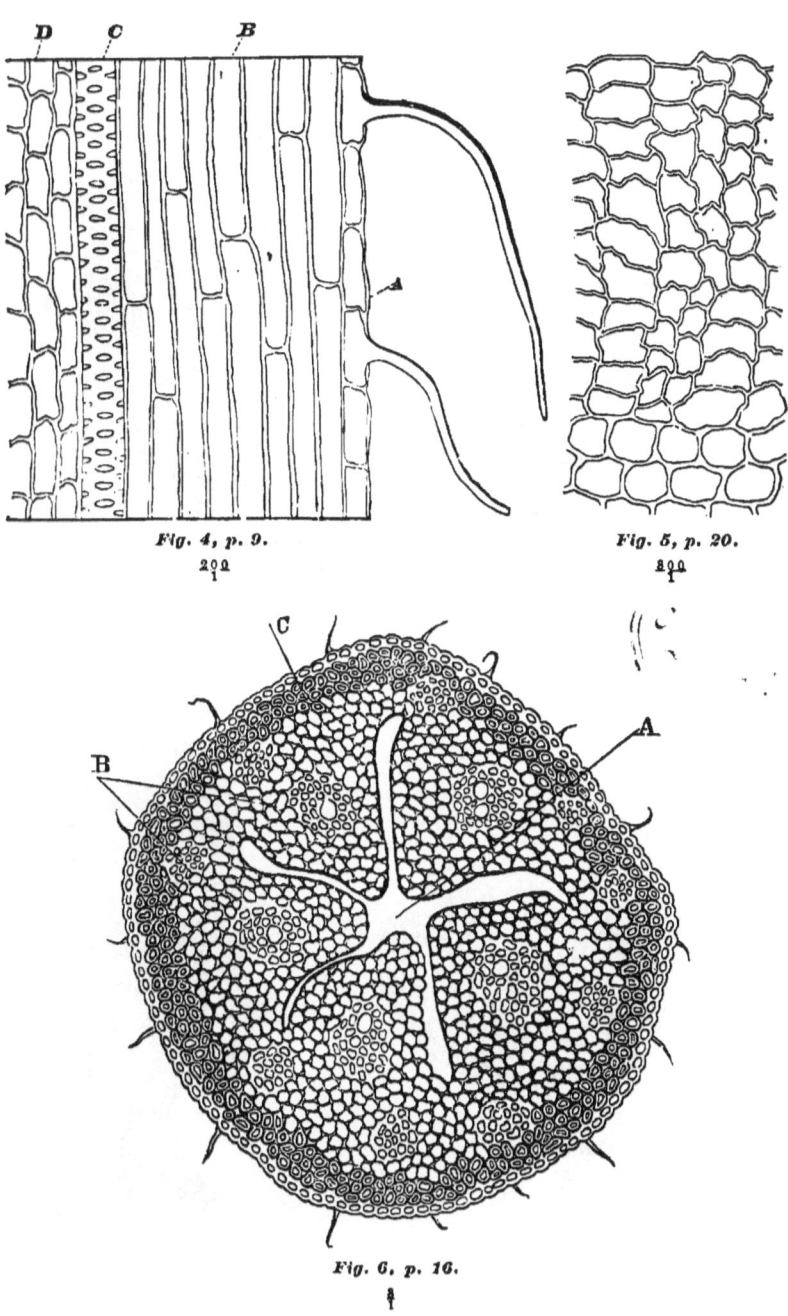

Fig. 4, p. 9. 200/1

Fig. 5, p. 20. 800/1

Fig. 6, p. 16. 3/1

100 PHENOMENA OF PLANT-LIFE.

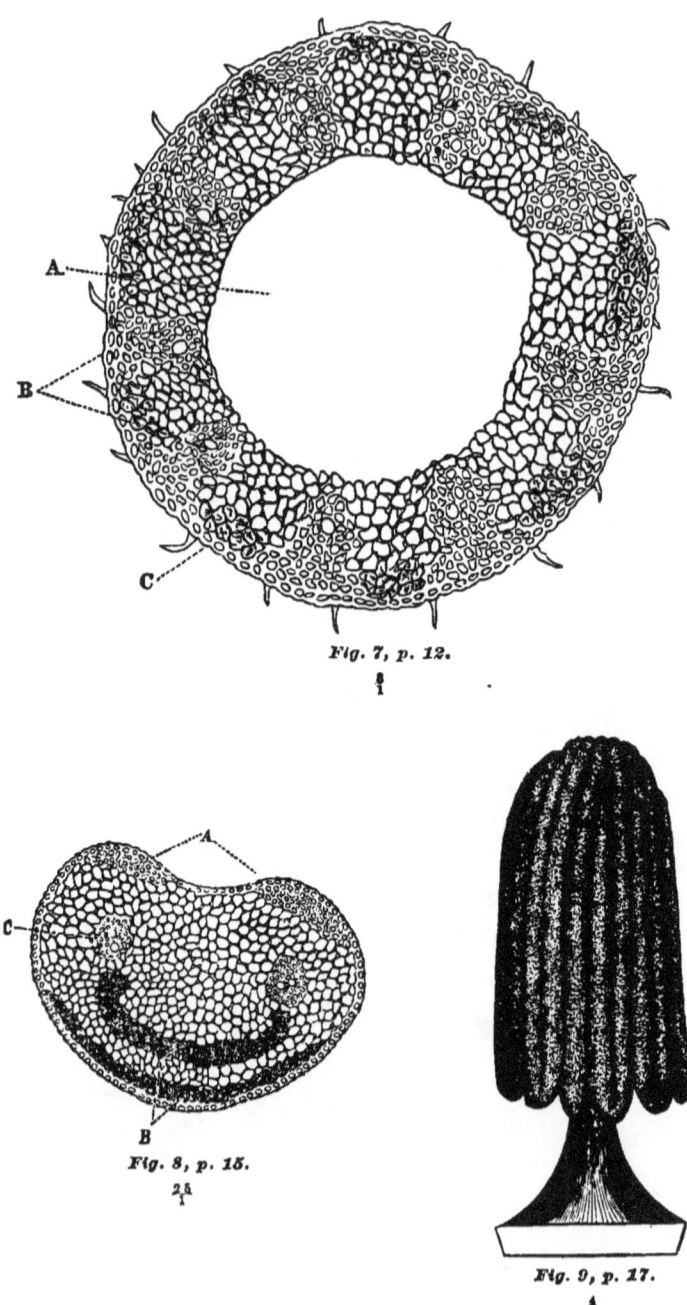

Fig. 7, p. 12.

Fig. 8, p. 15.

Fig. 9, p. 17.

PHENOMENA OF PLANT-LIFE.

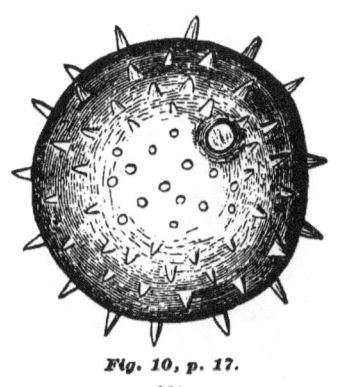

Fig. 10, p. 17.
¹⁵⁰⁄₁

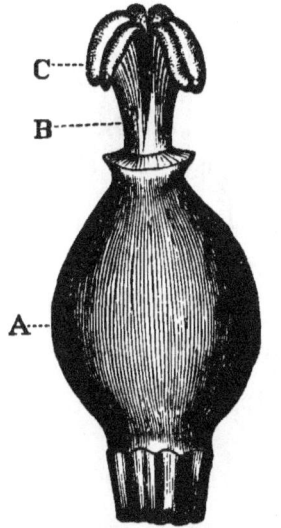

Fig. 11, p. 17.
⅔

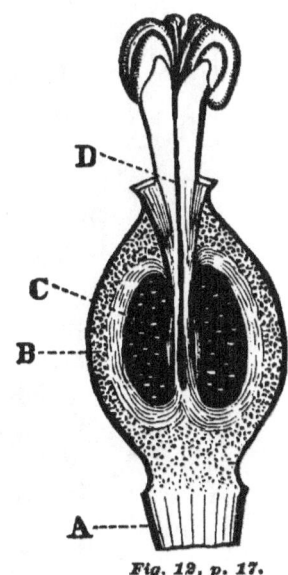

Fig. 12, p. 17.
⅔

Fig. 13, p. 17.
⅓

102   PHENOMENA OF PLANT-LIFE.

Fig. 14; p. 18.

PHENOMENA OF PLANT-LIFE. 103

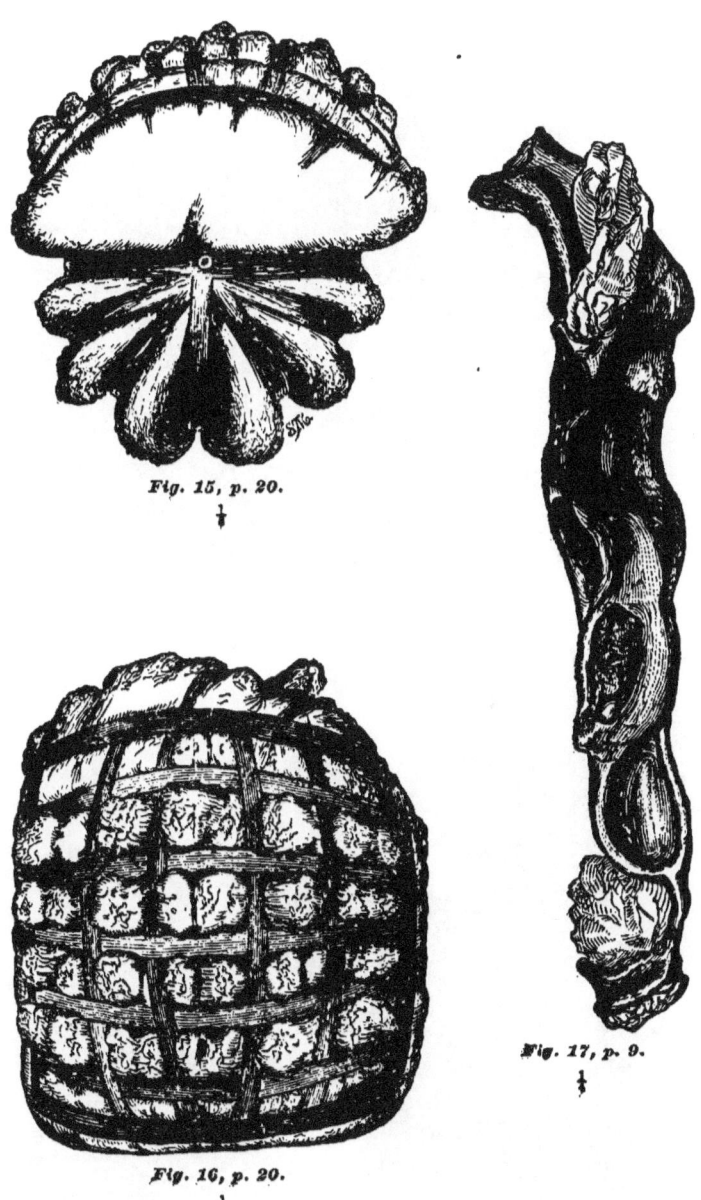

Fig. 15, p. 20.

Fig. 16, p. 20.

Fig. 17, p. 9.

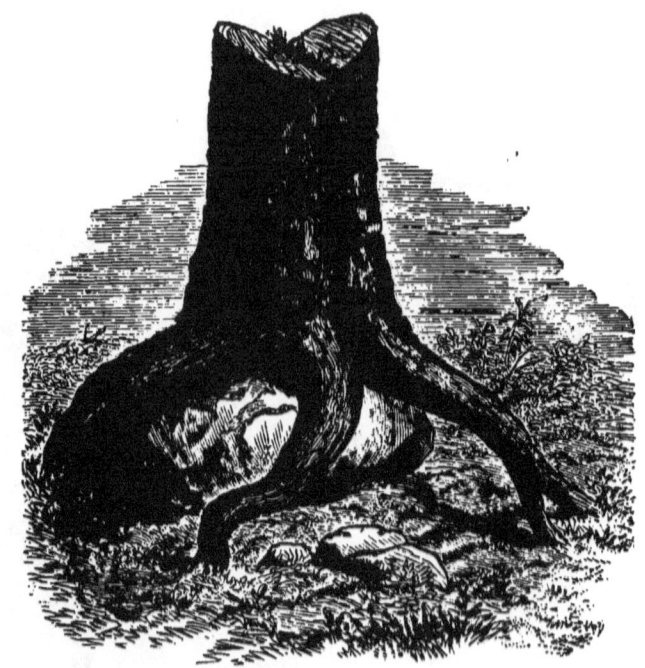

Fig. 18, p. 22.
1/20

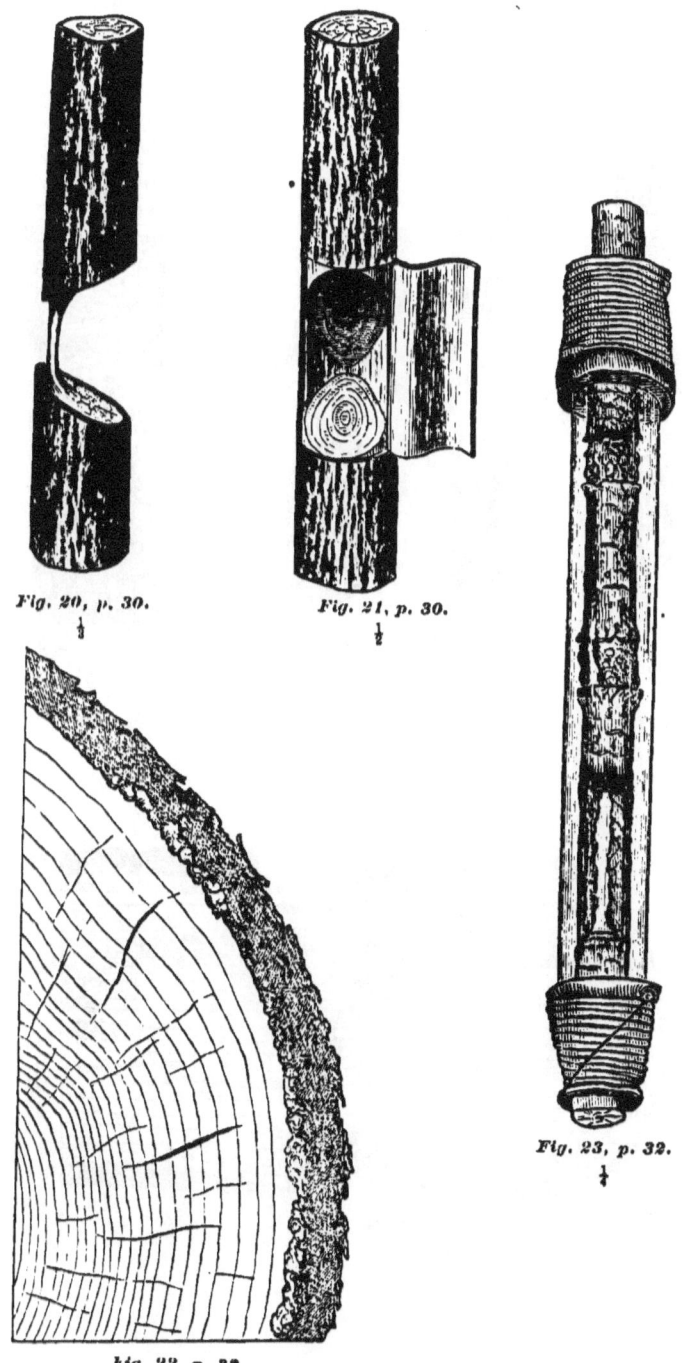

Fig. 20, p. 30.
Fig. 21, p. 30.
Fig. 22, p. 32.
Fig. 23, p. 32.

106   PHENOMENA OF PLANT-LIFE.

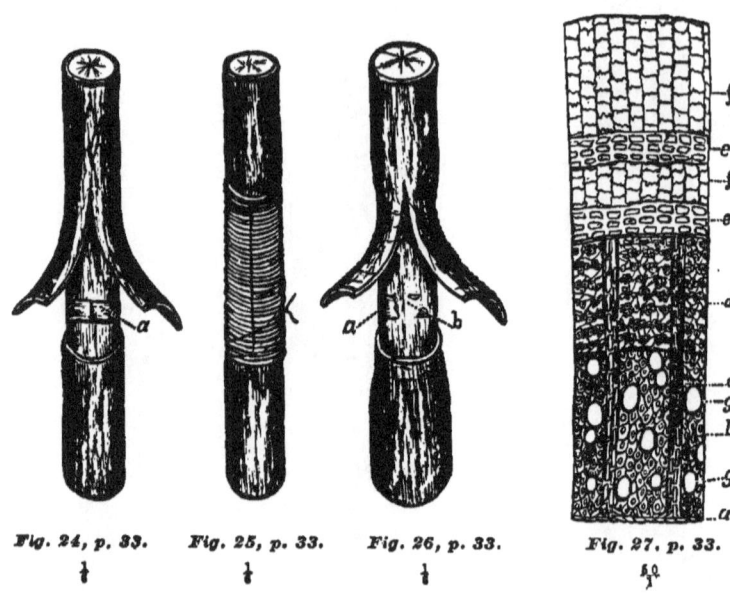

Fig. 24, p. 33.   Fig. 25, p. 33.   Fig. 26, p. 33.   Fig. 27, p. 33.

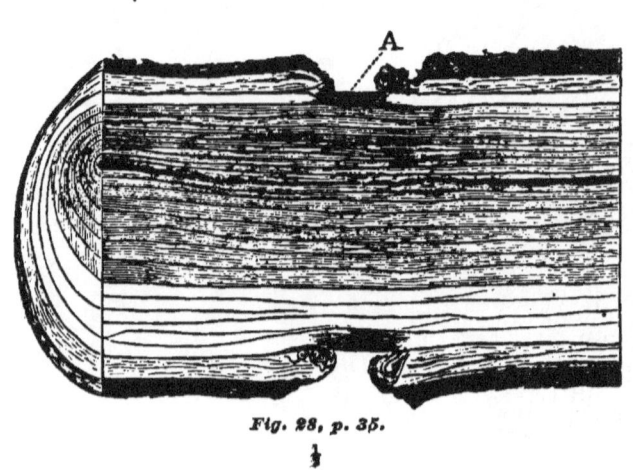

Fig. 28, p. 35.

PHENOMENA OF PLANT-LIFE. 107

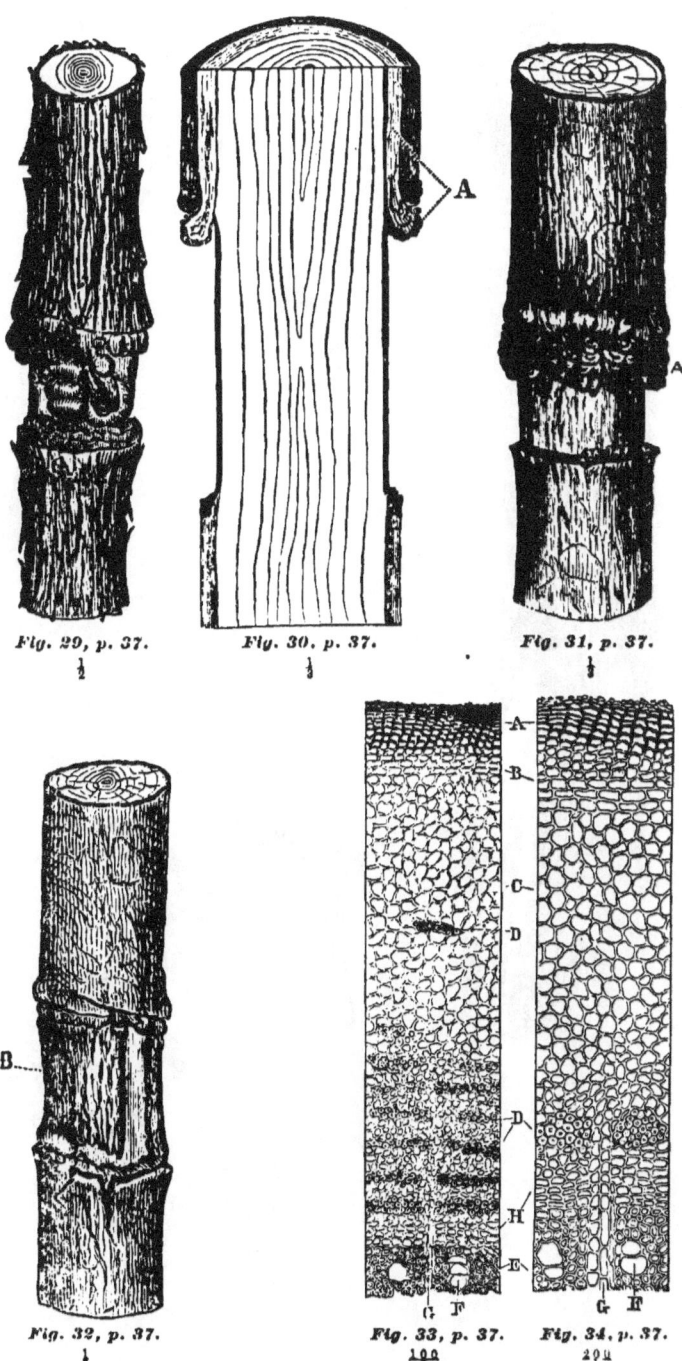

Fig. 29, p. 37. ½
Fig. 30, p. 37. ⅓
Fig. 31, p. 37. ⅓
Fig. 32, p. 37. ⅓
Fig. 33, p. 37. ¹⁰⁰⁄₁
Fig. 34, p. 37. ²⁰⁰⁄₁

108　PHENOMENA OF PLANT-LIFE.

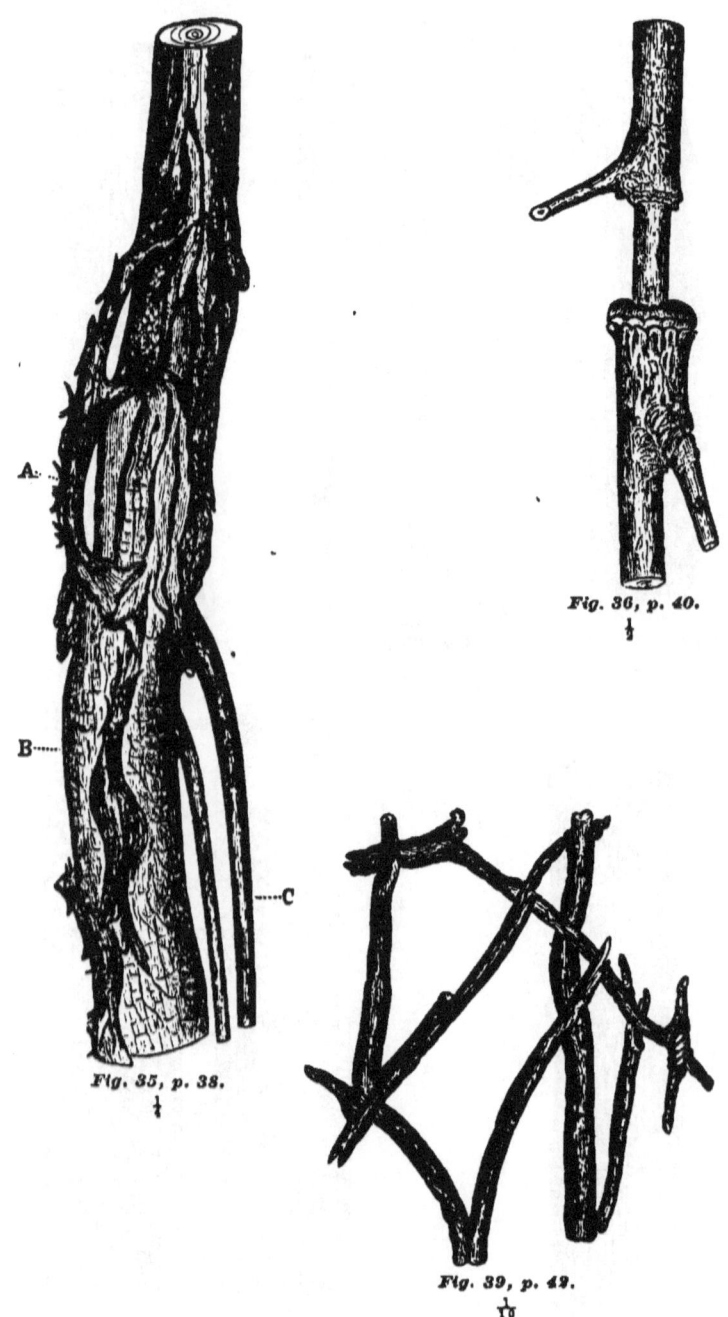

Fig. 35, p. 38.
¼

Fig. 36, p. 40.
½

Fig. 39, p. 42.
1/10

PHENOMENA OF PLANT-LIFE. 109

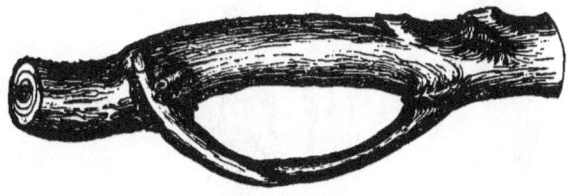

Fig. 37, p. 41.
½

Fig. 40, p. 42.
¼

14

110 PHENOMENA OF PLANT-LIFE.

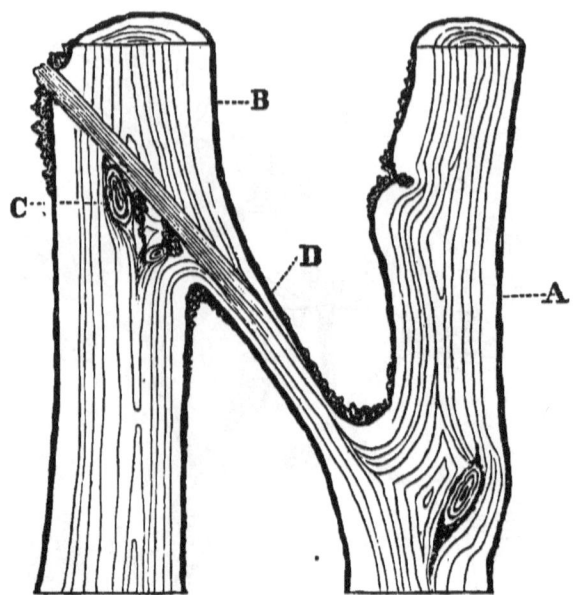

Fig. 38, p. 42.

Fig. 41, p. 42.   Fig. 42, p. 42.   Fig. 43, p. 43.

PHENOMENA OF PLANT-LIFE.

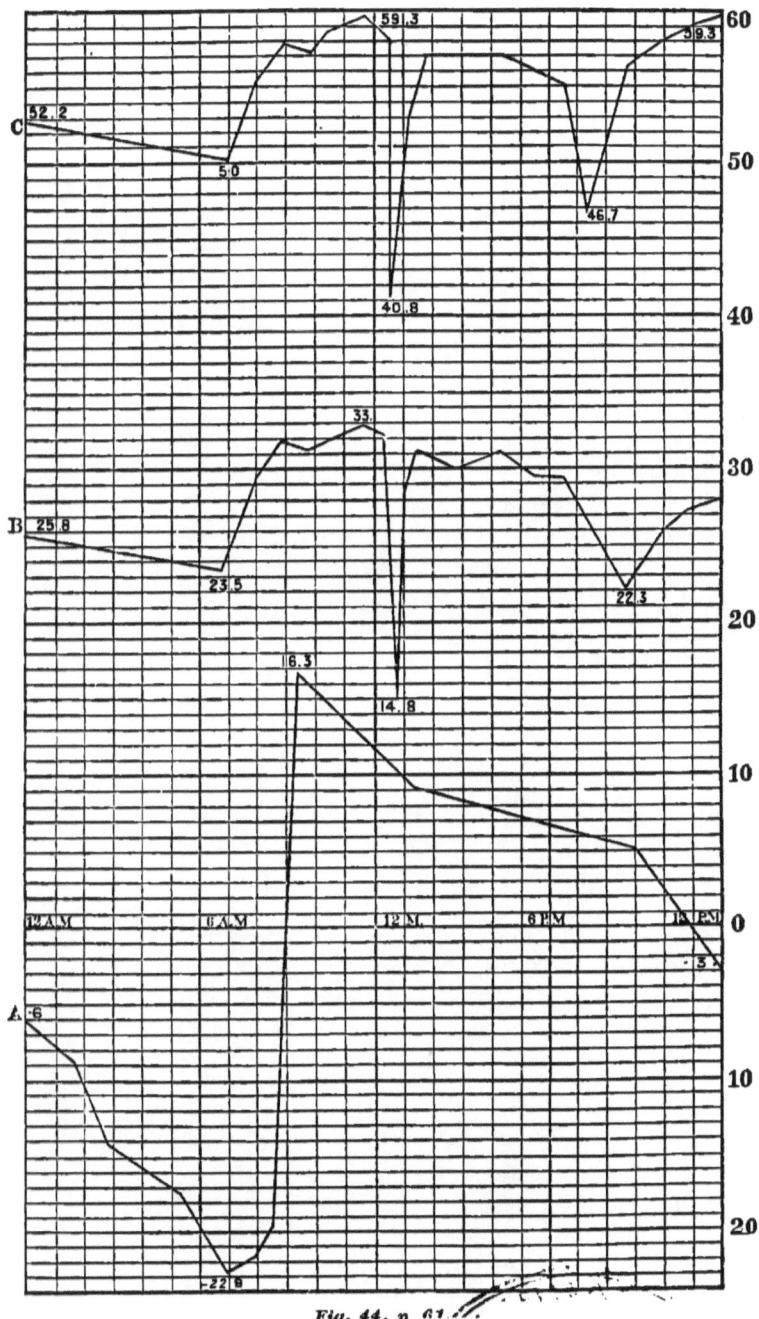

Fig. 44, p. 61.

www.ingramcontent.com/pod-product-compliance
Lightning Source LLC
Chambersburg PA
CBHW021945160426
43195CB00011B/1225